WITHDRAWN

D0276282
* 000242602 *

PROBLEM-BASED METHODOLOGY

RESEARCH FOR THE IMPROVEMENT OF PRACTICE

Related book of interest

EVERS & LAKOMSKI
Knowing Educational Administration: Contemporary Methodological Controversies in Educational Administration Research

PROBLEM-BASED METHODOLOGY

RESEARCH FOR THE IMPROVEMENT OF PRACTICE

by

VIVIANE ROBINSON

The University of Auckland, New Zealand

PERGAMON PRESS

OXFORD · NEW YORK · SEOUL · TOKYO

UK Pergamon Press Ltd, Headington Hill Hall,
Oxford OX3 0BW, England

USA Pergamon Press Inc, 660 White Plains Road, Tarrytown,
New York 10591-5153, U.S.A.

KOREA Pergamon Press Korea, KPO Box 315, Seoul 110-603,
Korea

JAPAN Pergamon Press Japan, Tsunashima Building Annex,
3-20-12 Yushima, Bunkyo-ku, Tokyo 113, Japan

Copyright © 1993 Pergamon Press Ltd

All Rights Reserved. No part of this publication may be reproduced, stored in a retrieval system or transmitted in any form or by any means: electronic, electrostatic, magnetic tape, mechanical, photocopying, recording or otherwise, without permission in writing from the publishers.

First edition 1993

Library of Congress Cataloging-in-Publication Data

Robinson, V. M. (Viviane M.)
Problem-based methodology: research for the improvement of practice/by Viviane Robinson.
p. cm.
Includes index.
1. Action research in education—Methodology. I. Title.
LB1028.24.R63 1993 370'.78—dc20 92-36833

British Library Cataloguing in Publication Data

A catalogue record for this book is available from the British Library

ISBN 0-08-041925-9

Transferred to Digital Printing 2005

Contents

Preface

Whether addressing educational practitioners or academics about the contribution of research to the improvement of practice, I have found that the following introductory dedication strikes a responsive and humorous chord in my audience:

> This address is dedicated to all those educational researchers who have wished that educational practitioners and policy-makers would take more notice of their work, and to all those practitioners and policy-makers who have wondered when educational researchers would produce something that was worth taking notice of.

The dedication captures the problem I grapple with in this book. While much educational research is relevant to practice, its impact seems to be much less than either researchers, practitioners or policy-makers would wish. This book presents an argument about why educational research has not made a greater impact and about how that impact can be increased.

In brief, I argue that educational researchers who want to have an impact on practice should adopt a problem-based methodology (PBM), in which the theories of action relevant to the problem situation are investigated, evaluated and, if necessary, altered. A theory of action is a theory we attribute to ourselves or others that purports to explain or predict, on the basis of relevant values, beliefs and motives why people act as they do in a given situation. To understand an educational problem, therefore, is to understand the theories of action of relevant agents and the factors that sustain those theories. To resolve an educational problem is to change those theories of action to produce consequences that are no longer judged to be problematic. Researchers must conduct these processes of problem understanding and resolution as a critical dialogue with practitioners, so that competing theories of the problem can be adjudicated and new theories of action learned during the course of the research itself, rather than left to some subsequent process of dissemination. Elaboration and defence of this argument for PBM requires a theory of problems of practice, a theory of how such problems are solved by practitioners via the construction of a theory of action, and a theory of how those solutions may themselves become problematic. In addition, it requires defence of a set of criteria by which solution adequacy is judged and a theory and practice of critical dialogue. It is this combination of methodological resources that constitutes problem-based methodology.

In Part I PBM is elaborated and defended through theoretical exposition; in Part II it is illustrated and reflected upon, via two original case studies conducted in high schools; in Part III it is systematically compared with three existing traditions of educational and social research. Part I begins with a discussion of how the contribution of research may be assessed and how its impact, or lack of it, is explained. I argue that there are important methodological reasons for the limited impact of educational research, including the mismatch between much of social science methodology and generic features of educational problems, such as their complexity, their value-ladenness and their intimate links to the theories of action of those involved in the problem situation. Research which is better matched to practice requires a theory of its subject matter, and this is developed through discussion of the psychological and philosophical literature on ill-structured problems. In addition, criteria of theoretical adequacy are proposed which are appropriate to the problem understanding and problem resolution purposes of PBM. Part I concludes with a theory of critical dialogue and discussion of how it can overcome the dilemmas involved in doing research which is simultaneously critical and collaborative.

Part II illustrates and reflects upon the methodology via two original case studies conducted over a three-year period in two high schools by a colleague and myself. The first of the case studies involves the problems of low impact and staff resistance in programmes of professional development and appraisal. The second involves the problem of making democratic management work in the face of such constraints as school size and school politics. The problem analyses show both what sustains each problem and why administrators have so far been unable to resolve them. The intervention process includes the feedback and negotiation of these analyses and the steps taken to overcome the problems. I discuss why the intervention was more successful in one case than the other, how disagreements were resolved through critical dialogue, the obstacles to such dialogue, and how PBM differs from other qualitative methodologies on issues of data analysis, validation and generalisation.

Part III further develops PBM by showing how it both builds on and departs from the empiricist, interpretive and critical traditions of educational research. The departures are justified in terms of the match of the methodology to its subject matter and to its goals of problem understanding and resolution. Comparison between PBM and empiricist methodologies requires a careful distinction between positivist and non-positivist empiricisms. I note that while there are considerable epistemological similarities between PBM and non-positivist empiricisms, their procedural differences, such as the emphasis in the former on holistic and accessible analyses and on collaborative and critical social relations of inquiry, should make them differentially effective for the resolution of problems of practice. While both PBM and interpretive inquiry emphasise practitioner understandings, the impact of the latter is diminished by its relativist epistemology and its non-interventionist stance. Critical research poses the greatest challenge to PBM because it shares the same practical purposes and proceeds through an apparently similar sequence of problem analysis, critique, education and transformative action. Despite these similarities, I argue that the practical contribution of the latter has been diminished by its failure to connect its analyses to the practices of particular agents who are, or can become, responsible for local problems; by its failure, in many cases, to move beyond critique, and by its exclusion of the powerful from its problem resolution processes. On the other hand, Habermas's writing on rational consensus has much in common with critical dialogue, and this commonality should be exploited in future empirical research on the micro-politics of problem resolution.

Given the plethora of new approaches to social and educational research, I need to be clear about precisely what is and is not being claimed for PBM. First, PBM is appropriate when researchers intend their work to contribute to the improvement of problems of practice. Since these are not the only purposes of educational research, I am not claiming that PBM is always the appropriate methodological choice. A knowledge of PBM, however, should help practitioners, researchers and those who fund them to understand why much research has had disappointing practical results, and what is required given that one does wish to make a practical contribution. The choice between PBM and other approaches is neither a choice between theory and practice, nor a choice between rigour and relevance. Research on problems of practice and on problems that arise within academic disciplines both make heavy theoretical and empirical demands. The choice that is involved is whether or not to engage the theorising of those involved in the problem situation. This does not mean privileging their perspective; it means being prepared to move back and forth between academic theory and theories of action in a mutually educative dialogue about the implications of each for the resolution of the selected problem.

Second, I do not claim that PBM is a paradigm, both because there is considerable doubt about the validity of such distinctions and because features of each so-called paradigm are to be found within it. Those readers whose methodological reference points and inspiration are defined by paradigm membership will, I hope, be challenged by this book to think across those boundaries and to evaluate their current research practice, not in terms of its adherence to paradigm characteristics, but in terms of its degree of inclusion of those features of inquiry which enhance their chances of making a contribution to the resolution of educational problems.

During the five or more years it has taken me to write this book and to complete the associated empirical work, I have despaired at times at the range of theoretical and practical expertise that is ideally required to answer the questions I have posed. At other times, however, I realised that if I had become too immersed in any one of the relevant specialities, I would never have been able to adopt the interdisciplinary approach that was required to formulate PBM. The Acknowledgements page indicates just some of the people who have helped me to move beyond my own disciplinary background as an organisational psychologist, to learn what was required from other disciplines. In addition, I myself could not have written this book without the immersion in the world of practice that came from my experience as an organisational consultant and as an academic deeply involved in graduate professional education. The practice context has always provided an important standard against which the significance of my research questions, and my own and others' answers, have been evaluated. It is this commitment to connecting the worlds of research and practice, in fact to seeing them as qualitatively indistinct, that was the driving force behind this book.

Acknowledgements

A book that has been as long in the making as this one is inevitably the product of numerous intellectual influences. My principal debt is to Chris Argyris, not only for his writing on methodology and intervention, but for his superb example, in both his teaching and research, of commitment to scholarly inquiry that informs practice. Drafts of portions of this book benefited greatly from the discussions that followed presentations at the University of Auckland, Monash University, Deakin University, Stanford University and the Oxford Polytechnic. In addition, presentations to the New Zealand, Australian and American Associations for Research in Education led to reworking of papers that contributed to this volume.

A special debt is owed the Deans of the Graduate Schools of Education at Monash and Stanford, for giving me the opportunity to spend a highly productive sabbatical in their institutions. At Monash, I am especially grateful to Colin Evers whose generosity with his time and philosophical expertise helped me to understand some of the epistemological underpinnings of my work. Other Monash colleagues whom I would like to acknowledge are Judith Chapman, Peter Gronn and Vicki Lee. At Stanford, I learned much from Denis Phillips's ability to clearly communicate complex ideas in the philosophy of social science. Others who have provided crucial intellectual and personal encouragement are Gary Anderson, Marie Clay, Brian Haig, Gabriel Lakomski, Patti Lather, Diana McLain Smith, Ken Strike, Christine Swanton, Helen Timperley and Jim Walker.

This book would not have been possible without the courage and commitment to learning of the principals and staff of Northern Grammar and Western College. I am also extremely grateful to Michael Absolum, my colleague in the case study research reported in Part II of the book. Thanks are also due to the Ministry of Education for funding the original research and for its openness to methodological diversity. The skills of Brenda Liddiard-Laurent were invaluable in final manuscript preparation. Finally, I cannot thank David enough, for his own superb skills in problem analysis and problem-solving were invaluable in bringing this project to fruition.

The following editors and publishers are thanked for permission to use previously published material.

Robinson, V. M. J. (1989a). Some limitations of systemic adaptation: The implementation of Reading Recovery. *New Zealand Journal of Educational Studies*, **24** (1), 35–45.

Robinson, V. M. J. (1989b). The nature and conduct of a critical dialogue. *New Zealand Journal of Educational Studies*, **24** (2), 175–187.

Robinson, V. M. J. (1992a). Doing critical social science: Dilemmas of control. *International Journal of Qualitative Studies in Education*, **5** (4), 345–359.

Robinson, V. M. J. (1992b). Why doesn't educational research solve educational problems? *Educational Philosophy and Theory*, **24** (2), 8–28.

Robinson, V. M. J. (in press). Current controversies in action research. *Public Administration Quarterly*.

Robinson, V. M. J. & Absolum, M. (1990). Leadership style and organisational problems: The case of a professional development programme. In N. Jones & N. Frederickson (Eds.), *Refocusing educational psychology*, pp. 31–54. London: Falmer Press.

Swanton, C. H. M., Robinson, V. M. J. & Crosthwaite, J. (1989). Treating women as sex objects. *Journal of Social Philosophy*, **20** (3), 5–20.

PART I

Justifying and Explaining Problem-based Methodology

1

The Need for a Problem-based Methodology

Every year, the American Educational Research Association, comprising some 17,000 scholars and practitioners from all over the world, awards a prize for a programme of research of high quality and practical significance. In 1990 the Raymond B. Cattell Early Career Award was given to Professor Jeannie Oakes, for her programme of research on school tracking, which she defines as a sorting process "whereby students are divided into categories so that they can be assigned in groups to various kinds of classes" (Oakes, 1985, p. 3).

In the address which she gave on receiving the award, Professor Oakes (Oakes, 1991) moved beyond a summary of the research on tracking, to address the question, "Can tracking research influence school practice?" She had turned to this issue because she felt increasingly committed to trying to answer the questions of policy-makers and practitioners about how to overcome the problems that the tracking research had highlighted. Professor Oakes explained her reactions to these questions as follows:

> These are very legitimate questions, and I find them troubling because responses such as, "Well, these matters extend beyond the scope of my research", while true, seem neither satisfying nor responsible. Therefore, my attention has turned increasingly to whether our tracking research is really able to inform practice, and since the answer for me seems, at best, "Only partly", I've started thinking more seriously about what type of research might actually help policy-makers and practitioners, who are interested in developing alternatives aimed at solving tracking-related problems (Oakes, 1991).

Why is it that this body of research can, at best, "only partly" inform practice? Oakes's evaluation seems puzzling, given the clear relevance of the research and her claim that it was mostly of good quality and remarkably consistent in its findings. In her address, Oakes offered several reasons for the limited practical contribution of research on tracking.

First, the problems which are pursued in this research are different in kind from the problems which are faced by educational practitioners. Tracking researchers have asked four different types of question: "What are the consequences of being in a particular track, for academic achievement, attitudes, peer relations and life chances?" "What are the social correlates of membership in particular tracks?" "What factors influence track placement?" "What are the effects of various track assignments on students' school and classroom experience?" Practitioners, in contrast, focus on the quite different problem of how to organise students of varying interest and abilities in ways that they believe to be educationally efficient and effective. Their problem is a practical one about what to do (Gauthier, 1963); the researchers' questions are about what is the case. Tracking is the practitioners' solution to their practical problem, and while they may acknowledge that solution has some or all of the unintended consequences that researchers have identified, they will not abandon it until they have another which satisfies the practical, normative and political constraints which they perceive to be relevant. Researchers have little to offer by way of alternative solutions, when the problems they have been studying are not those of the practitioner.

Second, as Oakes points out, tracking researchers have been far more concerned with describing the problem, or at least their version of it, than with developing alternative practices, and this failure is a second reason why tracking research has, as yet, little to offer to the practitioner.

The third reason why tracking research fails to connect with practice, or at least with its reform, is that it lacks a theory of change with which to address the complexities involved. These complexities, according to Oakes (1991), are technical, normative and political. At the technical level, researchers need to address questions such as how to teach heterogeneous groups, without leaving the slower students behind, or forcing the quicker ones to wait. The history of reform attempts teaches us, however, that technical innovations are frequently insufficient for problem resolution. Unless teachers and administrators feel committed to such changes, the old problems will resurface despite new organisational forms. Such commitment will not develop unless we change many deep-seated beliefs about individual difference, about race and class correlates of achievement, and about efficient forms of pedagogy, because while these are in place, detracking reforms will simply not make sense to many of those involved. In addition to these normative considerations, the political issues must be addressed, such as the opposition which comes from those who are currently advantaged by tracking arrangements.

This discussion of tracking research and its connection to practice introduces several key themes of this book. One theme concerns the nature of practitioners' problems, and why doing research that connects with practice involves understanding how practitioners attempt to resolve them. A second theme concerns the normative basis for evaluating current practice and for advocating the desirability of alternative ways of understanding and resolving educational problems. A third theme concerns the theory and practice of change that will help researchers to move beyond problem description to collaborative problem resolution. Before developing these themes any further, however, Professor Oakes's question about the practical contribution of research on tracking needs to be set in the wider context of the debate about the contribution of educational research, in general, to educational practice.

The Disputed Contribution of Educational Research to Practice

Sensible debate about the extent of the contribution of research to practice requires clarification of the nature of practice, and of what counts as a contribution.

What is Educational Practice?

Following Carr and Kemmis (1986, p. 81), I define educational practice as action informed by beliefs about how to achieve educationally important purposes in particular circumstances. This definition incorporates both the delivery of educational services and the formulation of policies intended to influence those services. Given the tenuous connection between policy and service delivery, however, this book is concerned more with the influence of research on the latter than on the former. This is not to deny that in particular cases changes in policy are a necessary precursor of improved services. A direct focus on services, however, enables one to test rather than assume the connection between these two levels of practice.

Practices are more than behaviours, since they incorporate beliefs about both what is important and about how what is important can be realised in particular circumstances. Since the reasoning processes that inform practice include both technical considerations about how to achieve goals and normative considerations about what one's goals should be, research may contribute to practice by improving either the technical or normative reasoning of practitioners. By practitioners I mean anyone who engages in educational practices, including students, parents, teachers, administrators and policy-makers. Finally, practices are enacted by individuals, groups, organisations and even cultures. The study and improvement of educational practices, therefore, can take place at any or all of these levels of analysis, depending on the scope of the practices under investigation.

What Counts as a Contribution?

The debate about the contribution of educational research to educational practice reflects widely varying views about both the extent of the contribution and about the appropriateness of the expectation itself. Robert Travers represents one of the more positive participants in the debate. In his book *How research has changed American schools* (Travers, 1984) he describes how research on intelligence testing, on behavioural approaches to teaching and learning and on child development has had a lasting and widespread impact on what happens in American classrooms. The positive conclusions of Travers contrast sharply with those of Phillips (1980), who answers the question "What do the researcher and the practitioner have to offer each other?" with the rather pessimistic conclusion "not much". He writes, "By expecting little from each other, occasionally both the researcher and the practitioner may be in for a small treat" (p. 20).

The debate continued in the 1985 *World Yearbook of Education* which was devoted to the relationship between research, policy and practice. Mitchell (1985), in his chapter on the impact of research, acknowledged that widely varying conclusions had been drawn about its contribution:

> Both enthusiastic support and deep scepticism have been widely expressed regarding the quality of recent educational research activity and its utility in the development of educational policy and practice (p. 20).

While one could conclude from all this that perhaps more research is needed on the impact of educational research, a more useful conclusion is probably that a clear verdict is impossible because the question is badly conceived. Some writers, such as Travers, equate the contribution of research with the promotion of change in educational policy or practice, and base their positive judgements on this criterion. Writers like Phillips (1980), Kohn (1976) and Cuban (1990), however, equate contribution, not with change, but with the far more demanding criterion of the improvement of educational and social practice, frequently interpreted as the resolution of educational and social problems. Cuban, for example, shows how research on teacher-centred and student-centred instruction, on academic and vocational curricula, and on centralised and decentralised forms of school governance, have fed repeated cycles of educational change without altering the educational problems that triggered the reform attempts in the first place. Questions about the contribution of educational research to educational change are very different from questions about the contribution of educational research to the resolution of problems, and nothing follows from the answers to the former about the answers to the latter.

Despite the logic of the distinction between change and improvement, one could argue that since most writers judge the changes they write about to be improvements, the distinction is unnecessary for the purpose of assessing the contribution of educational research. This equation of change with improvement is possibly defensible when we have good evidence for, and agreement on, the desirability of particular innovations, but less justifiable when the evidence is equivocal or its interpretation contested. There is certainly controversy, for example, about whether or not the introduction of intelligence testing constituted an improvement over previous allocation practices. Despite this, Robert Travers (1984) implicitly equates change with improvement throughout his book on the impact of research, and thus avoids the important normative issue of how we judge the desirability of particular changes.

The criterion of improvement rather than change is adopted in this book, because gaining agreement about the worth of current and alternative educational practices is seen as a major challenge in the process of connecting research and practice. In addition, educational research has a potential for either improving or worsening the material and mental condition of those it seeks to influence (Leboyer, 1988; Mitchell, 1985), and a criterion of improvement rather than of change forces researchers and their audiences to deal with difficult normative questions about the purposes and persons served by their research activity.

How is Improvement Judged?

To announce an interest in the improvement of educational practice is to immediately invite questions about how it is to be judged. I adopt a pragmatist view of improvement, whereby practice is said to be improved when problems that arise in the pursuit of our goals or the satisfaction of our needs are resolved in ways that enhance our ability to resolve other problems that we experience (Walker, 1987, pp. 4–5). This criterion implies that short-term expediency in the solution of one problem does not constitute improvement, because such solutions make the resolution of other problems more difficult. In short, a solution to any one problem must cohere with solutions to our whole problem set. The extremely difficult issues involved in further defining what counts as a problem and as a solution are introduced below and then discussed at greater length in Chapter 2.

Educational problems, in the sense of problematic practices, are identified by a discrepancy between actual and desired states of affairs. (A second more neutral sense of "problem" is introduced in Chapter 2.) The desired state serves as the standard against which a situation is described as problematic, and as having improved. Professor Oakes's (1991) conclusion to her summary of the research on tracking illustrates the process. "To the extent that we value the quality of what students experience in school as an end in itself, these inequalities in opportunities and experience become a critical schooling problem." She suggests that current practice falls short of standards of equality, and signals through the use of the conditional ("to the extent that") the need to check the degree of agreement about the standard itself.

In talking about solutions to educational problems one must not overlook the fact that different groups may employ different solution criteria, and that such judgements may change over time as beliefs shift about what is educationally desirable and possible. Perhaps the concept of resolving educational problems captures this iterative quality better than the more definitive concept of a solution. Uncertainty and controversy over these judgements does not imply, however, that agreement cannot be reached and that any such judgement reflects no more than the relative power of various interest groups. Disputes about what counts as a problem may be rationally resolved by linking the disputed practice to another whose value is not in dispute. For example, if relevant practitioners disagree about whether the high failure rate of minority students is a problem, the disagreement may be resolved by showing how reduction of these rates is linked to their shared desire for a classroom environment that is safe and conducive to learning. Assuming for the moment that standards of solution adequacy can be agreed to, I am suggesting that educational researchers who wish to make a contribution to practice adopt the goal of problem understanding and resolution, rather than the goal of change. Precisely what is involved in doing this is the subject matter of this book.

Achieving Change or Resolving Problems? A Research Example

It is time to illustrate some of the implications of the distinction between the resolution of educational problems and the achievement of educational change for the assessment of the contribution of educational research. The Reading Recovery programme, which is based on decades of New Zealand research on early reading processes, has been implemented throughout New Zealand primary schools and in several United States, United Kingdom and Australian school systems (Clay, 1987, 1990).

Reading Recovery is a programme of early intervention which provides individualised reading and writing instruction to those children who have the lowest level of reading achievement after one year of school. The programme has changed the way children are assessed and the way those who are identified as low progress readers are subsequently assisted. These changes in teaching practices are in turn supported by policy changes in the way resources such as teachers, supervisors and training opportunities are allocated. There is no doubt that the research associated with the Reading Recovery programme (Clay, 1985) has led to major changes in educational practice and policy at the junior school level. In New Zealand, a national cadre of Reading Recovery teachers and tutors is in place, along with an administrative structure to finance and maintain the quality of the programme. In addition, the vast majority of those who enter Reading Recovery improve their reading levels while in the programme.

Given my previous distinction between the achievement of educational change and the resolution of educational problems, however, we must also ask whether the latter, as well as the former, has been achieved. The answer will depend in part on the way we define the problems which the research and the subsequent Reading Recovery programme were designed to address. One problem can be seen as the low reading levels of those children who enter the programme. Given an expectation of progress for every normal child, such failure constitutes an educational problem, whose resolution is judged by evaluating whether the children improve in the programme and sustain their gains when they return to regular classroom instruction. This problem definition was accepted by the developers and administrators of the programme, and much of the evaluation effort went into monitoring the progress of individual children, both during and after their period of individualised remedial instruction.

A second problem can be seen in the failure of the regular teaching programme to cater for all beginning readers. Evidence for the improvement of this problem would be judged by a decline in the rate of referral to the programme, attributable to improved classroom instruction in beginning reading. In contrast to the first problem, there was disagreement about whether or not the failure of the regular programme to cater for all children did constitute an educational problem. The programme developers argued that it was inevitable that some children would fail under regular instruction and, therefore, that the problem was one of the identification and recovery of failing readers, rather than the prevention of such failure in the first place (Robinson, 1989a).

An independent evaluation of the New Zealand programme showed that while children made substantial gains under individual instruction, their rate of progress slowed considerably in the regular class environment once the tuition was completed. On follow-up, their reading level was only slightly ahead of a comparison group which had not entered the programme (Glynn, Crooks, Bethune, Ballard & Smith, 1989). While it was hoped that each child would be operating as a "self-improving system" by the end of the programme, the skills they learned were typically not powerful enough to overcome any limitations in the contexts (both home and school) in which they were expected to further their development. On this problem, therefore, we have evidence of research contributing to substantial educational change without resolving the relevant educational problem.

Given the contentious nature of the second problem, that of preventing reading failure, it is not surprising that there is little evidence available on which to make a judgement. It is important for the purposes of this discussion, however, to examine the possibility that the dispute could, in principle, be resolved by examining the implications of the contentious practices for the achievement of those that were not in dispute.

In the case of Reading Recovery, there is agreement that the failure of normal children to learn to read is an educational problem, but disagreement about whether or not the failure to teach such children to read in the regular classroom also constitutes a problem. After all, it may be inevitable that a certain percentage of children fail under any given system of instruction. The disagreement could be resolved by showing how resolution of the first agreed problem of individual reading failure is linked to the second contested problem of systemic teaching failure. The independent evaluation suggested that the two were closely linked, because long-term solutions to individuals' reading failure required change to some classroom instructional practices. In other words, acceptance of the second problem may be a necessary condition for resolution of the first.

In summary, the Reading Recovery example illustrates the difference between judging the impact of research on the basis of change and judging it on the basis of problem resolution. While the latter criterion invites far more disputes about the extent of the contribution, researchers can contribute to the resolution of such disputes by tracing the implications of contested practices for the achievement of less contested values and goals.

Why isn't the Contribution Greater?

Two types of explanation have been offered for the limited impact of educational research upon practice. The first type attributes the difficulties to the processes of dissemination and utilisation that are used to try and connect research to the context of practice (Hoyle, 1985). The second type attributes the difficulties to features of the research itself, in particular to the way the practical problem is transformed in order to meet standards of rigour, and to the way that practitioners' theories of action are bypassed (Argyris, Putnam & McLain Smith, 1985; Mitchell, 1985).

Explanations Related to Research Dissemination

The literature on the dissemination and utilisation of educational and social research proceeds on the assumption that it is the role of researchers to produce reliable generalisations which can be disseminated to those who can apply them in the service of improved educational and social practices (Keeves, 1988). One explanation for the shortfall in expected contribution points to the failure of educational research to deliver the goods which can then be applied. Phillips (1980), for example, in an article on the contribution of research to educational practice, concludes that: "The bottom line is that social scientists have not been able to discover generalisations that are reliable enough, and about which there is enough professional consensus, to form the basis for social policy" (p. 17). He argues that one reason for this apparent failure lies in the complexity of the phenomena that social scientists investigate. One aspect of this complexity is that individual differences in human abilities and aptitudes lead people to react differently to different "treatments". Phillips quotes Cronbach in arguing that unravelling these interactions requires very complex designs and huge data sets. In addition, the resulting generalisations prove unstable as the characteristics of the particular groups studied change over time. Chester Finn, Assistant Secretary of Education under Reagan, also sees educational research as having difficulty in producing generalisable knowledge, but, unlike Phillips, attributes the difficulty to shortfalls in the educational research community itself, rather than to the complexity of the subject matter (Finn, 1988).

Most writers who accept the dissemination model, however, believe that there are ample goods to be delivered; the problem lies in the process of dissemination and utilisation of research. This view has led to an effort to discover the characteristics of innovations, of adopters and of dissemination processes, which are associated with the utilisation of research and its resulting innovations (Cousins & Leithwood, 1986; Dunn, Holzner & Zaltman, 1988; Keeves, 1988). Cousins and Leithwood's (1986) review of sixty-five evaluation studies conducted between 1971 and 1985 sought to identify features of the evaluations and of the utilisation settings that were associated with research utilisation. They found that evaluations were more likely to be used when the methodology was appropriate to the decision requirements, when the users were involved in the evaluation, when the findings were consistent with users' beliefs and expectations, and when there was a minimum amount of conflicting information available from other sources.

Factors associated with successful dissemination and adoption of psychosocial interventions (Backer, Liberman & Kuehnel, 1986) are remarkably similar to those identified by Cousins and Leithwood. Once an intervention is validated by clinical and field trials, and associated with a technology that makes it accessible to practitioners, then critical factors in adoption relate to the degree of interaction between the sponsors and the adopters, and the assistance received with the process of change, from both within and outside the adopting organisation.

In summary, these findings suggest that adoption is more likely when there is a match, or when a match is created through intensive interaction, between the theoretical and value framework of the research and of the context of application. This theme is found in Cousins and Leithwood's (1986, p. 360) conclusion that evaluation findings were more likely to be used when "they were consistent with the beliefs and expectations of the users" and in Nisbet and Broadfoot's claim (1980, pp. 53–59) that research has impact when it fits the way the problem is being framed at the time.

One might conclude from all this that practice-oriented researchers should increase the chances of utilisation by developing theories and recommendations that are consistent with the practice culture. There are two major problems with this advice. First, the mismatch between the research and practice culture frequently arises because standards of methodological rigour make it impossible to preserve the complexity of the practice situation. Following this advice sets up a dilemma between practical relevance and methodological rigour. Second, the advice suffers from the failure to make the distinction, defended earlier, between the achievement of change and the resolution of problems. While a match will increase the probability of change, it will decrease the chance of problem resolution when that requires challenging key features of the practice culture. Both these issues are concerned with the match between the way researchers and practitioners seek to understand and resolve problems. They invite examination of the way in which research methodology itself influences the contribution made by educational research.

Explanations Related to Educational Research Methodology

A second type of explanation for the limited contribution of educational research focuses on features of the methodology typically employed in research which seeks generalisable findings. By methodology I mean "the study—the description, the explanation and the justification—of methods, and not the methods themselves" (Kaplan, 1964, p. 18). Methodology is a meta-level investigation of the limitations, resources and presuppositions of methods, aimed at understanding the process of inquiry rather than the products themselves (Kaplan, 1964, p. 23). In focusing on methodological explanations of the research–practice gap, I am concerned with the practical consequences of the choices researchers make about which methods to employ.

Different problems and different theories

One methodological explanation focuses on the differences between the types of problems faced by practitioners and researchers and the types of theories they employ to solve them. The problems practitioners face are problems about what to do, and what counts as a solution is constrained both internally by their own beliefs and values, and externally by material conditions and institutional and cultural expectations. This constraint structure, together with the strategies that satisfy it, constitute a theory of how to understand and resolve the problem. To return to the tracking example, teachers solve the problem of how to assign students to teachers by specifying material and ideological conditions which any solution must satisfy. The material conditions might relate to limited numbers of teachers and teaching spaces; the ideological conditions might specify what counts as appropriate pedagogy, given certain beliefs about the needs of students of different abilities. The theories which practitioners employ to understand and resolve such practical problems are called theories of action (Argyris & Schön, 1974, pp. 4–6).

Researchers also address and solve problems by the development or employment of theories, but they are of a significantly different type from those relevant to practitioners. The problems that researchers address gain significance from theories that are current within the relevant research communities, rather than within practical contexts, and the solutions are thereby subject to a different set of constraints from those that operate on the practitioner. The answers of the latter must take into account all those factors believed to be relevant to effective action in the practice context. The answers of the researcher are constrained by theoretically derived hypotheses, and factors which interfere with the derivation of such answers must be eliminated as far as possible. Since these conditions can rarely be obtained in the natural setting, researchers frequently transform the domain they are investigating through the use of various experimental and statistical control procedures. While such a transformation increases the chances that the evidence is interpretable within the researcher's theory, it results in a loss of meaning for the practitioner, since the variables controlled out by the researcher may be those that are most significant to practitioners' decisions about how to act. Hackman (1985), for example, in discussing research on incentive schemes, argues that worker reactions to such schemes are dependent on their social background, the presence of unions and the past history of union–management relations, and that these are precisely the factors that are controlled out of laboratory studies of incentive systems.

The transformation of the practice context in order to derive relatively unambiguous answers to the researcher's questions, poses major problems for the applicability of knowledge derived from this type of research. Practitioners are likely to judge research findings and associated theories as oversimplified, if the variables to which they are most responsive are omitted, or as unusable if causal theories highlight variables over which they have no control.

The consequences of bypassing theories of action

When researchers transform the context of practice, and in so doing bypass the theories of action that practitioners employ in the practice situation, the applicability of the researcher's theory is frequently questioned. Another consequence of this bypassing is the impoverishment of our causal theories of the problem. Knowledge of theories of practice gives us access to the reasons why people act to create the situations which are called problematic. Admittedly, there are major methodological problems in gaining access to these reasons, but assuming these can be overcome, theories of action tell us the meanings, values and purposes behind people's actions, and enable us to judge the extent to which they are implicated in the problem situation. If theories of action are causally implicated, the non-coercive resolution of such problems requires the development of new theories of action which those involved judge to be superior to those they employ at present. A change process which proceeds as if the practitioner either operated without a theory, or as if any such theory could be safely ignored, runs the risk of being controlling, ineffective or both. When there are significant differences between the theories of the researcher and the practitioner, the latter's theories are a source of objections to the relevance and applicability of the research. If these theories are bypassed, the objections appear as *ad hoc*, defensive reactions rather than as the reflection of an implicit rival theory. Researchers who seek to improve practice need a methodology which links such objections to the theory which gave rise to them, and which involves practitioners in adjudicating between the now explicit rival theories of the problem.

Reducing the Gap Between Research and Practice

Given the causal role of theories of action in educational problems and their role in determining the acceptance of research findings, it makes sense for those researchers who wish to improve practice to investigate rather than bypass these theories. This book develops, defends and illustrates such a methodology. The purpose of problem-based methodology (PBM) is to contribute to the understanding and improvement of problems of practice. In brief, PBM involves the reconstruction of theories of action which are operative in the problem situation, the evaluation of such theories, including the assessment of their possible causal role in the problem, and, where necessary, the development, implementation and evaluation of an alternative theory of action. Ideally, these stages of inquiry are embedded in a "critical dialogue" between researcher and practitioner; that is a conversation that is simultaneously critical and collaborative (Robinson, 1989b). I say "ideally", because several sections of this book (Chapters 3, 8 and 11) are devoted to discussion of some of the psychological and sociological barriers to such a dialogue.

This call for educational researchers to pay more attention to theories of action is not new, so some initial comparative exposition will help to clarify the nature of PBM. Carr and Kemmis (1986), for example, also believe that the research–practice gap is attributable, at least in part, to the way researchers bypass theories of action.

> The gaps between theory and practice which everyone deplores are actually endemic to the view that educational theory can be produced from within theoretical and practical contexts different from the theoretical and practical context within which it is supposed to apply. Consequently, because this sort of view is so widespread, it is hardly surprising that the gaps thereby created are interpreted as impediments that can only be removed by finding ways of inducing teachers to accept and apply some theory other than the one they already hold. If, however, it is recognised that there is nothing to which the phrase "educational theory" can

> coherently refer, other than the theory that actually guides educational practices, then it becomes apparent that a theoretical activity explicitly concerned to influence educational practice can only do so by influencing the theoretical framework in terms of which these practices are made intelligible (p. 115).

Carr and Kemmis are claiming that the gap between research and practice is caused by the mistaken belief that one can do educational research without investigating the theories that actually inform practice. Their advice for closing the gap is that we accept that all educational research is about theories of practice, and that the only way to change practice is through changing such theories.

While I agree with Carr and Kemmis's call for researchers to focus on the theories that inform practice, I do not agree that such theories are the only ones that qualify as educational. To equate educational theories with practitioners' theories is to tie research too tightly to the investigation of the status quo. One of the roles of educational researchers is to show how current practice reflects particular conceptions of education, and how different conceptions may lead to more valuable educational processes and outcomes. If educational research is restricted to the investigation of theories of action, the result may be neglect of the type of theorising required to critically evaluate that practice, and such critique is an essential component of PBM. For example, researchers who study children's learning outside formal schooling give us a picture of what children can do outside the provisions which we currently call educational. Their work gives us a basis for rethinking and reforming these provisions. Unlike Carr and Kemmis (1986), I believe that the gap between research and practice is not caused by the development of theories that lie outside practice, but by the failure to connect any such theories with those that are currently operative in the practice situation.

If PBM is not the only way to do educational research, is it the only way to do research that makes a difference to practice? Again, I disagree with Carr and Kemmis's somewhat sweeping answer: "... a theoretical activity explicitly concerned to influence educational practice can only do so by influencing the theoretical framework in terms of which these practices are made intelligible" (p. 115). Researchers can, in alliance with relevant educational and governmental agencies, induce or coerce compliance with practices and policies which are in conflict with practitioners' current theories of the situation. The busing saga in Boston during the late 1970s is a good example of a major educational change forced through by the power of judicial and governmental processes against the wishes of the majority of those directly involved. Carr and Kemmis's point should be that changing theories of practice is a necessary condition of non-coercive change, rather than of change in general. This is the position taken in PBM.

The methodological implications of the claim that non-coercive change requires change to theories of action are far from obvious. Does PBM methodology require researchers to actively engage with practitioners in a mutually educative way? How appropriate are non-interactive influence processes, such as talks, journal articles and the dissemination of scholarly advice? The answer to this question has been foreshadowed in the previous discussion of research dissemination and utilisation. It depends on the type of change that is required to solve the problem. Some problems can be solved by making changes which do not challenge core values and beliefs about the nature of the problem and how to solve it. For example, teacher stress may be resolvable by giving teachers skills in time management and relaxation techniques. Such technical adjustments have been variously called single-loop learning (Argyris & Schön, 1974, pp. 18–19) and first-order change (Watzlawick, Weakland & Fisch, 1974). Since these changes are compatible with existing ways of thinking, intensive intervention is not required. In many cases, however, problems cannot be resolved without challenging and changing core values and beliefs. The teachers' stress may persist despite workshops on time management and relaxation, because it is due to characteristics of their job design and school organisation. When resolution requires major changes to the assumptive framework that practitioners bring to the problem situation, double-loop or second-order change is needed (Argyris & Schön, 1974, pp. 18–19; Watzlawick, Weakland & Fisch, 1974). The literature on research utilisation suggests that this type of change is unlikely to be triggered by a methodology which bypasses the theory of the practitioner, develops a substantially different theory, and does not explicitly examine the implications of each for the maintenance and resolution of the problem situation.

PBM investigates both types of problem by reconstructing the theory or theories of action that are operative in the problem situation, evaluating how they may be causally implicated, and, if necessary, offering an alternative. When resolution of a problem requires double-loop change, and therefore more intensive types of intervention than are offered in the traditional approach to dissemination, PBM is particularly appropriate because of its inclusion of a model of critical dialogue, designed to resolve disagreements about the relative merits of researchers' and practitioners' theories, and to help the latter adopt an agreed alternative.

Some researchers may protest at the assumption, fundamental to PBM, that it is possible to reduce the research–practice gap by making changes to research methodology. Phillips (1980), for example, sees the research–practice gap as a consequence of the inevitable difference between the two activities. Relevant research findings, he argues, are only one of many factors that practitioners weigh up in deciding what to do. Financial constraints, political considerations and levels of skill and commitment may exert a pull in the decision process that is far stronger than relevant research findings. Rather than adjust our methodology, Phillips would have us adjust our expectations of the contribution of research.

My view would be that the current differences between the contexts of research and practice are constructed by the way we do research, and that the costs of those differences, for our ability to connect with practice, are too high. When researchers decide to investigate an educational problem via academic theories and to bypass theories of action, they are making a methodological choice that has significant implications for the likely impact of the research. If researchers routinely inquired into the theories of action that were operative in the situation they sought to influence, they would learn, at least, how central or peripheral their variables of interest were to the reasoning of those they sought to influence. Where academic theories and theories of action are significantly different, the choice to bypass the latter has additional methodological implications, because it reduces theory competition and thus weakens the theory-testing process. These methodological choices create rather than reflect differences between the research and practice contexts.

In summary, this book will argue that the gap between research and practice can be reduced, in the service of understanding and resolving educational problems, by reconstructing, evaluating and, if necessary, changing theories of action. There are several advantages to such a focus. First, such an approach enriches our understanding of educational problems, because theories of action tell us why people in the problem situation act as they do. Second, it makes explicit and adjudicates the theoretical differences between the researcher and practitioner which are the source of any mutual rejection. Third, if the practitioner's theory is shown to be implicated in the problem, then a non-coercive resolution of the problem requires development and testing of an alternative which is judged by practitioners to be both practical and more effective. PBM offers a process for engaging practitioners in making such judgements.

The Dilemma Between Acceptance and Effectiveness

The earlier discussion of the dissemination literature showed that research was more likely to have an impact on practice when its theoretical framework matched that employed in the practice situation.

The subsequent discussion of theories of practice, however, suggested that research which mirrors key features of practice is unlikely to be effective in resolving educational problems when those features are themselves implicated in the development and maintenance of the problem.

In such cases, the practice-oriented researcher is clearly in a dilemma. Recommendations that are consistent with current assumptions are likely to be acceptable but ineffective in resolving the problem. Recommendations which are incompatible with current assumptions are more likely to be effective, but are likely to be judged as unacceptable or irrelevant and thus never implemented. This dilemma is not a central one for educational researchers who do not expect or wish their analyses to influence practice, but for the rest, the dilemma is a major barrier to making a practical contribution. I return to the example of the Reading Recovery programme, because it illustrates the importance of this dilemma and the difference between the strategy employed by the researchers and that suggested by PBM. I previously argued that Reading Recovery research and the associated intervention had contributed to educational change without resolving the underlying problem of early reading failure. In the following analysis, I suggest that this limitation is due to the way the programme leaders responded to the dilemma between the acceptability and the effectiveness of their intervention.

The research on which the Reading Recovery programme was based suggested that many conditions in junior classes were not conducive to the early detection and recovery of failing readers. For example, whole class or even small group teaching made it difficult for teachers to make the sensitive and frequent observations needed to assist young readers in difficulty. Rather than challenge these practices, Reading Recovery personnel negotiated to set up alternative instructional programmes which incorporated the teaching practices which their research had shown to be essential to the swift recovery of low progress readers. An infrastructure of Reading Recovery teachers, teacher trainers, tutors and curricula supported the alternative instructional programmes (Robinson, 1989a).

The programme developers correctly judged that there were significant differences between the Reading Recovery and regular class programmes, and that the acceptance of the programme by the host systems would be jeopardised by attempts to modify regular classroom practices (Clay, 1987). They protected the theory of the research and bypassed the theories of the host systems by encapsulating the former within an alternative instructional system. This move was designed to ensure that the conditions necessary for effectiveness could be implemented, while reducing the likelihood that the host system would reject the innovation. One consequence of this move has been that the programme has been adopted and survived, even in host systems whose theories of reading instruction are discrepant with those central to the programme. The cost of not challenging these theories, however, has been that many of the children that succeed in the programme cannot maintain their gains in regular classroom environments where resources and instructional practices are not sufficiently supportive of their new skills (Glynn, Crooks, Bethune, Ballard & Smith, 1989). Bypassing theories of action operative in the regular junior classroom increased the acceptance but jeopardised the effectiveness of the programme.

The Reading Recovery example has shown that researchers cannot resolve problems by sacrificing effectiveness in the interests of acceptance, but neither can they resolve them by ignoring practitioners' negative reactions. In PBM, the acceptance–effectiveness dilemma is resolvable because both of these values are simultaneously and collaboratively pursued during the life of the research project. If the theories of practice of relevant practitioners are involved in the problem, then the methodology provides for collaborative examination of the merits of these theories and of the researcher's alternative. The ensuing dialogue is not a dissemination phase that takes place after the research is finished, but a collaborative process of theory appraisal that is central to the research itself (Hoyle, 1985).

Summary

Educational research has made a greater contribution to educational change than it has to the resolution of educational problems. A major reason for this difference is that many educational problems are irresolvable unless fundamental assumptions about the nature of the problem and its solution are altered. Educational researchers have typically targeted problematic practices and not the theories of action in which those practices are embedded, so new practices have typically been assimilated to old ways of thinking.

Problem-based methodology focuses on theories of action and how they might sustain the practices and outcomes that are considered to be problematic. This relationship is stated as a hypothesis, because not all educational problems can be explained in this way. For example, practitioners might correctly understand what is causing a problem and how to prevent it, and be willing to act accordingly, but be prevented from doing so by a shortage of resources. If causal relations are demonstrated between relevant theories of action and the educational problem, the next step, in PBM, is dialogue with relevant practitioners about the suggested causal links and the merits of an alternative theory of practice. This dialogue serves as a type of theory appraisal, since the practitioner's theory will be the source of objections to that of the researcher and vice versa. The acceptance–effectiveness dilemma can be resolved if objections to the research are treated as theory competition rather than as political obstacles that are irrelevant to the conduct of the inquiry itself.

The great majority of educational research, as I will show in Part III of this book, does not take theories of action nearly as seriously as does this suggested problem-based methodology. Instead of investigating theories of action and their implications for educational problems, researchers have typically developed and tested theories that gain significance from the relevant research community rather than from the practice context. If these theories omit variables which practitioners consider to be central to their practice, or if they highlight variables or causal relations which are not manipulable in the practice context, they will be judged as irrelevant by practitioners.

PBM does not replace research which bypasses theories of action, because the latter provides an essential critical resource for the understanding and resolution of educational problems. Such research will not contribute directly to the improvement of practice, however, until it is incorporated into a methodology which uncovers, evaluates and, if necessary, alters the theories that are currently operative in the problem situation.

References

Argyris, C. & Schön, D. (1974). *Theory in practice: Increasing professional effectiveness.* San Francisco: Jossey-Bass.

Argyris, C., Putnam, R. & McLain Smith, D. (1985). *Action science.* San Francisco: Jossey-Bass.

Backer, T. E., Liberman, R. P. & Kuehnel, T. G. (1986). Dissemination and adoption of innovative psychosocial interventions. *Journal of Consulting and Clinical Psychology,* **54**, 111–118.

Carr, W. & Kemmis, S. (1986). *Becoming critical: Education, knowledge and action research.* London: Falmer Press.

Clay, M. M. (1985). *The early detection of reading difficulties,* 3rd ed. Auckland: Heinemann.

Clay, M. M. (1987). Implementing Reading Recovery: Systemic adaptations to an educational innovation. *New Zealand Journal of Educational Studies,* **22** (1), 35–58.

Clay, M. M. (1990). The Reading Recovery programme, 1984–88. *New Zealand Journal of Educational Studies,* **25** (1), 61–70.

Cousins, J. B. & Leithwood, K. A. (1986). Current empirical research on evaluation utilization. *Review of Educational Research,* **56** (3), 331–364.

Cuban, L. (1990). Reforming again, again and again. *Educational Researcher,* **19** (1), 3–13.

Dunn, W., Holzner, B. & Zaltman, G. (1988). Knowledge utilization. In J. P. Keeves (Ed.), *Educational research, methodology and measurement: An international handbook,* pp. 220–226. Oxford: Pergamon Press.

Finn, C. E. (1988). What ails education research? *Educational Researcher*, **17** (1), 5–8.

Gauthier, D. P. (1963). *Practical reasoning*. London: Oxford University Press.

Glynn, T., Crooks, T., Bethune, N., Ballard, K. & Smith, J. (1989). *Reading recovery in context*. Wellington: Department of Education.

Hackman, J. R. (1985). Doing research that makes a difference. In E. E. Lawler, A. M. Mohrman, S. A. Mohrman, G. E. Ledford, & T. G. Cummings (Eds.), *Doing research that is useful for theory and practice*, pp. 126–175. San Francisco: Jossey-Bass.

Hoyle, E. (1985). Educational research: Dissemination, participation and negotiation. In J. Nisbet & S. Nisbet (Eds.), *Research policy and practice* (*World Yearbook of Education*), pp. 203–217. London: Kogan Page.

Kaplan, A. (1964). *The conduct of inquiry*. San Francisco: Chandler Publishing Company.

Keeves, J. P. (1988). Knowledge diffusion in education. In J. P. Keeves (Ed.), *Educational research, methodology and measurement: An international handbook*, pp. 211–219. Oxford: Pergamon Press.

Kohn, M. L. (1976). Looking back—A 25 year review and appraisal of social problems research. *Social Problems*, **24**, 94–112.

Leboyer, C. L. (1988). Success and failure in applying psychology. *American Psychologist*, **43** (10), 779–785.

Mitchell, D. E. (1985). Research impact on educational policy and practice in the U.S.A. In J. Nisbet & S. Nisbet (Eds.), *Research policy and practice* (*World Yearbook of Education*), pp. 19–41. London: Kogan Page.

Nisbet, J. & Broadfoot, P. (1980). *The impact of research on policy and practice in education*. Aberdeen: Aberdeen University Press.

Oakes, J. (1985). *Keeping track: How schools structure inequality*. New Haven: Yale University Press.

Oakes, J. (1991). Can tracking research influence school practice? 1990 Raymond B. Cattell Early Career Award Address, AERA 1991 Annual Meeting (Cassette Recording No. RA-1-7.14). Chicago: Teach 'em Inc.

Phillips, D. C. (1980). What do the researcher and practitioner have to offer each other? *Educational Researcher*, **9** (11), 17–24.

Robinson, V. M. J. (1989a). Some limitations of systemic adaptation: The implementation of Reading Recovery. *New Zealand Journal of Educational Studies*, **24** (1), 35–45.

Robinson, V. M. J. (1989b). The nature and conduct of a critical dialogue. *New Zealand Journal of Educational Studies*, **24** (2), 175–187.

Travers, R. M. W. (1984). *How research has changed American schools*. Kalamazoo, MI: Mythos Press.

Walker, J. C. (1987). Democracy and pragmatism in curriculum development. *Educational Philosophy and Theory*, **19** (2), 1–10.

Watzlawick, P., Weakland, J. & Fisch, R. (1974). *Change*. New York: Norton.

2

Taking Problems and Practitioners Seriously

Taking problems seriously involves doing research in a way that reflects a theory of the generic characteristics of problems; taking practitioners seriously involves studying the theories of action that they employ to solve problems. Both types of theory are included in the conceptual apparatus of PBM which is summarised in Figure 2.1. This chapter describes this conceptual framework by presenting a theory of a problem, a theory of how they are solved by theories of action, and a set of standards to judge solution adequacy. Finally, PBM incorporates a model of the social relations of inquiry which are conducive to the development of a shared understanding of problems and of how to solve them. This model, which is described as critical dialogue, is described and illustrated in Chapter 3.

A methodology ought to reflect, rather than be indifferent to, whatever knowledge we have about the subject of its inquiry. A theory of a problem has a central place in PBM because greater isomorphism between methodology and educational problems may reduce the problem of irrelevance besetting social research. In addition, since both researchers and practitioners need to employ, evaluate and revise theory in order to understand and resolve problems, much of what is said about the one group may be applicable to the other. If we understand how practitioners solve problems, and incorporate such understanding into our methodology, we may be far more successful than at present in making the connection between our research and their practice.

Two different literatures are relevant to investigating the nature and resolution of educational problems. The literature on ill-structured problems is useful because it analyses the cognitive demands presented by many educational situations and shows how practitioners theorise in order to both understand and solve the problems they present. The philosophical writing on ill-structured problems provides a conceptual analysis of the characteristics of such problems and of the cognitive

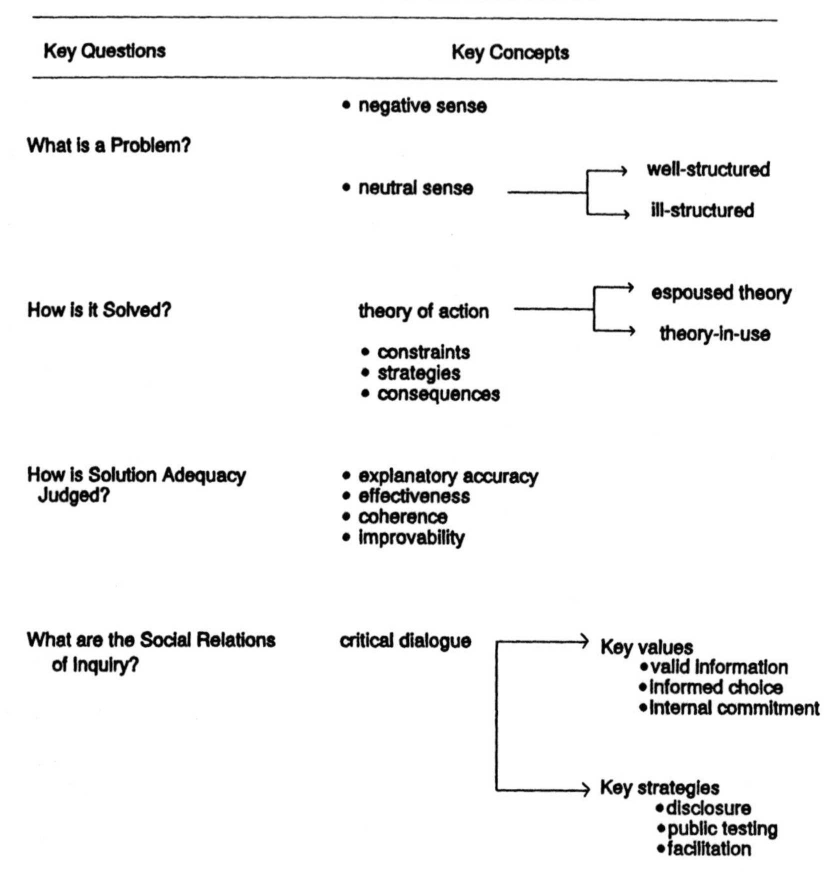

FIGURE 2.1. The conceptual framework of problem-based methodology.

demands they pose (Nickles, 1981; Simon, 1973). The psychological literature on ill-structured problems provides detailed case studies of practitioners' problem setting and solving strategies and comparisons of the approach taken by experts and novices (Chi, Feltovich & Glaser, 1981; Robinson & Halliday, 1988; Voss & Post, 1988).

The second literature which helps one to understand the nature of ill-structured problems and their resolution is richer than the former in providing a normative basis for evaluating the adequacy of the theories which actors construct to solve such problems. Argyris's work, for example, is centrally concerned with how theories of action which are designed to solve a problem can either fail to do so, or can, in so doing, create further problems (Argyris, Putnam & McLain Smith, 1985). A research methodology which takes practitioners seriously involves reconstructing and evaluating the theories of practice which are hypothesised to be implicated in the selected educational problem. In order to do this, researchers need some understanding of the general characteristics of the theories they are attempting to reconstruct, and of the problems which these theories were designed to solve.

Educational Problems as Ill-structured

A problem was defined in Chapter 1 as a gap between an existing and a desired state of affairs, because this is what is meant when we identify educational practices and their consequences as problematic. Low achievement levels of minority groups, sexist curricula, early reading failure and demoralised teaching staff are all examples of educational problems because they describe states that fall short of generally accepted standards for educational practice. This negative sense of a problem as something undesirable contrasts with the more neutral sense of a problem as a puzzle or challenge that needs to be solved. Practitioners who have to design new programmes, figure out how to cater for handicapped children, or allocate financial resources to different departments, are faced with educational problems in this second neutral sense. The starting point is not a problematic practice to be explained and eliminated, but a goal to be achieved.

Figure 2.2 shows the relationship between the negative and the neutral sense of "problem". When problems in the neutral sense of a goal to be achieved are inadequately solved, the result is a problem in the second negative sense of the word. The return arrow between inadequate solutions and goals to be achieved represents the way practitioners recycle through a problem-solving process until an adequate solution is found. It also represents the methodological requirement to locate problematic practices in the problem-solving processes that gave rise to them.

Since problems in the negative sense are inadequate solutions to problems in the neutral sense, there is considerable theoretical and practical value in knowing how the latter are solved, and how those processes may yield solutions which are themselves judged to be problems. Thomas Nickles's constraint inclusion account of problem-solving is useful for these purposes. He defines a problem, in the neutral sense, as "... a demand that a certain goal be achieved plus constraints on the manner in which the goal is achieved, i.e., conditions of adequacy on the problem solution" (p. 111). The process of problem-solving *is* the process of setting or discovering solution constraints. As Nickles (1981) puts it: "The more constraints on the problem solutions we know, and the more sharply they are formulated, the more sharply and completely we can formulate the problem, and the better we understand it" (p. 88). Practitioners draw on past experience, background theory and feedback from the environment to establish and

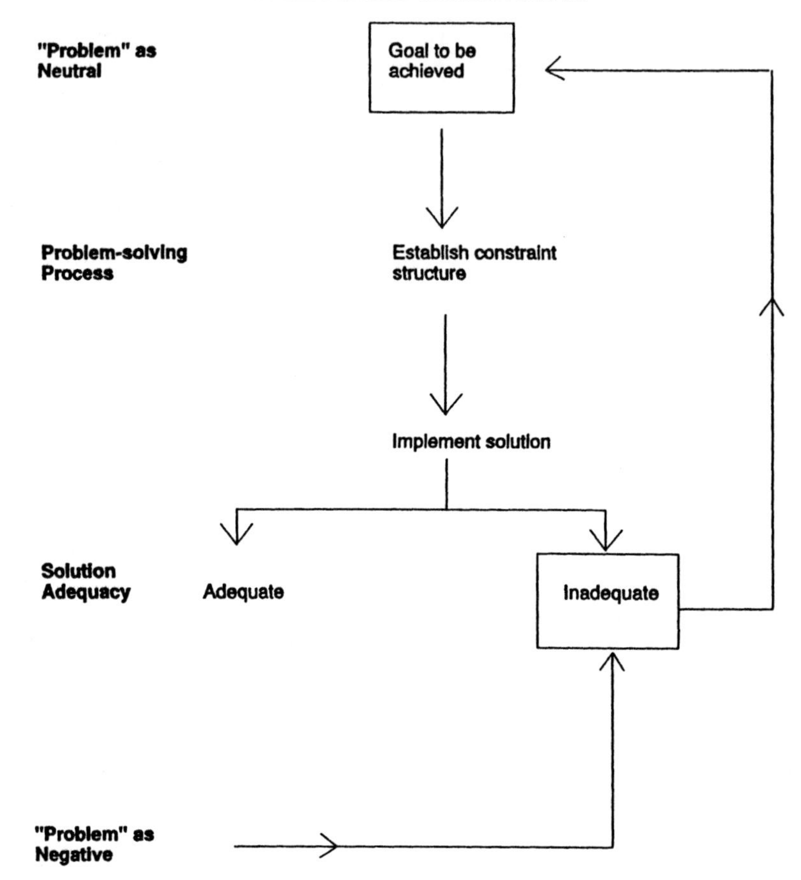

FIGURE 2.2. The relationship between the neutral and negative senses of "problem".

revise a constraint structure. Research comparing experts and novices shows that much of the advantage of the latter is due to their richer store of relevant knowledge and their ability to conceptualise it in ways that organise a problem solution (Voss & Post, 1988).

A distinction is usually made in the problem-solving literature between well-structured and ill-structured problems; a distinction which reflects the degree of specification of relevant constraints, rather than the quality of the constraints themselves. A problem is described as ill-structured when it lacks obvious criteria for solution adequacy, when the means for reaching a solution are unclear, and when there is uncertainty about the nature and availability of the required information. Well-structured problems, by contrast, have clear solution criteria, definable procedures for reaching the solution, and specifiable information requirements (Simon, 1973).

Most educational problems, in common with the vast majority of real world problems, are of the ill-structured variety (Frederiksen, 1984). For example, when educators are required to produce a curriculum plan there will be uncertainty and even conflict over what counts as a good one, an unlimited number of ways of producing it, and a huge and varied amount of information and resources that could be drawn upon in developing it. Similarly, the school counsellor and psychologist must help clients to solve problems which are open to a number of different interpretations and hence to a number of different solutions (Robinson & Halliday, 1988).

An ill-structured problem is unsolvable as such, because there are too many areas in which to search for information, and insufficient constraints on what counts as a solution. The goal statement "prepare a curriculum plan" specifies the end product, but the problem is not solvable until other constraints are set which specify what sort of product counts as adequate. These may concern the guiding educational philosophy of the school, limitations of resources such as money, equipment and staff, and the expectations of the local community and central authorities. The problem is now, not one of producing a plan, but of producing one which meets a number of constraints on solution adequacy. Simon argues that ill-structured problems like this one are solved by breaking them down into a series of smaller well-structured problems, which are co-ordinated through a design programme to create a coherent solution. This is how he summarises the way an architect solves the ill-structured problem of building a house:

> The whole design, then, begins to acquire structure by being decomposed into various problems of component design, and by evoking, as the design progresses, all kinds of requirements to be applied in testing the design of its components. During any given short period of time, the architect will find himself working on a problem which, perhaps beginning in an ill-structured state, soon converts itself from evocation from memory into a well-structured problem. We can make here the same comment we made about playing a chess game: the problem is well-structured in the small, but ill-structured in the large (Simon, 1973, p. 190).

If ill-structured problems are solved by transforming them into well-structured ones, then the assignment of a problem to one or other of these categories must depend, at least in part, on where the solver is in the solution process. The same point can be made from the perspective of the relevant community of problem solvers; if the community agrees on a particular formulation of a problem, it will be seen as well-structured within that community (Voss & Post, 1988). This suggests that educational problems present as ill-structured either because they are under-theorised, or because there is disagreement about the adequacy of any of the rival solution candidates.

At this point, it is worth pausing to consider some possible objections to the relevance of this account of problem-solving to the educational researcher. One could question the accuracy of portraying practitioners as struggling with uncertainty, and enmeshed in problems for which they creatively design and test possible solutions. Maybe the school leader reached for last year's curriculum plan and made a few alterations before presenting it to the school board. Maybe the school counsellor and psychologist use a few well-developed routines for understanding and treating the vast majority of their clients' problems. Louis Pondy and Anne Huff, in a study of the decision-making of school administrators, found that they accomplished even quite major changes through the application of well-established administrative routines, rather than through a more dramatic rethinking of existing concepts and procedures (Pondy & Huff, 1985).

The discrepancy between the preceding account of ill-structured problem-solving and evidence that practitioners' work is largely the application of pre-established routines can be resolved by treating the routines as the *results* of prior problem-solving efforts. Whether these routines were developed by the practitioners who now apply them, or by others, they represent the transformation of ill-structured problems into well-structured ones, and they will be applied routinely and with little cognitive effort. The transformations will be accepted and their origins forgotten while their adequacy as solutions to prior problems is accepted. On occasion, however, shifts in the task environment, in the knowledge of those involved, or in the composition of the relevant community, trigger a re-evaluation of solution adequacy. What was once a solution is now seen as a problem and the process of setting new constraints on solution adequacy begins again. Nickles (1981, p. 114) captures the process with his quip "... what, in turn, is a theory but a problem's way of generating new problems?!".

Theories of Action as Problem Solutions

Problems are resolved by developing a constraint structure which establishes the parameters of acceptable solutions. The constraints, together with the strategies which satisfy them, constitute the theory of action for the problem situation. From the point of view of the actor, a theory of action is a theory of control, because it specifies how to achieve particular purposes under given conditions. From the point of view of an observer, a theory of action has explanatory and predictive power, because it tells us why actors behave as they do, and how they are likely to behave in future (Argyris & Schön, 1974, p. 6).

Two sorts of theories of action are distinguished on the basis of the type of evidence from which they are derived (Argyris, Putnam & McLain Smith, 1985, pp. 80–82). Theories of action which are inferred from how people say they behave, or would behave, are called espoused theories. These tell us the constraints that actors seek to satisfy, or believe they have satisfied, in their problem-solving efforts. Theories of action which are inferred from actual behaviour are called theories-in-use, and they tell us the constraints which actually governed their problem-solving. The degree of congruence between an espoused theory and theory of action is a matter for empirical inquiry in every case; the accuracy of people's self-reports is likely to vary across type of subject matter and conditions of reporting.

It was argued in Chapter 1 that if researchers investigate problematic practices and outcomes in the context of the problem-solving processes that gave rise to them, they will understand the reasoning that sustains the problem and that may need to be altered in order for it to be resolved. In the subsequent section of this chapter, the relationship between ill-structured problems, the theories of action that constitute their solution, and the way those solutions can in turn produce further problems is illustrated via a detailed example of ill-structured problem-solving.

An Example of Ill-structured Problem-solving

The example involves the practical problem of how to give negative feedback to a colleague who is perceived to be performing poorly. There is considerable evidence, from both business and educational organisations, that this type of problem is inadequately solved, in that far from improving performance, negative feedback often results in lowered motivation, little improvement and a worsened relationship between the participants (Bridges, 1986; Cusella, 1987). Feedback senders either fail to deliver a clear and honest message, or if they do, they are punitive in the process. In other words, this research suggests that practitioners frequently solve the neutral problem of how to give negative feedback, in ways that produce problems, in the negative sense, of lowered motivation and little improvement in performance. In the rest of this chapter I show how PBM can help us understand why this is the case, by reconstructing and evaluating the theories of action that practitioners use to solve the problem of how to give negative feedback. I start by describing a negative feedback situation and explaining why it constitutes an ill-structured problem. Next, I describe how a particular practitioner solves the problem by presenting, in dialogue form, the data from which his/her theory of action will be inferred. Those readers who are unfamiliar with dialogue analysis may wish to follow the argument first, and then check it against the data during a second reading. The subsequent section shows how inferences are drawn from the dialogue to construct the theory of action of the practitioner. Finally, I tackle the question of the adequacy of the theory by describing and applying four criteria of theory appraisal.

The negative feedback situation that features in this example was developed as part of a programme of research designed to understand the theories actors employ in this situation and the reasons for their ineffectiveness (Argyris, 1982). It requires respondents to give feedback to an imaginary colleague (Ted) about the way he had conducted an interview with one of his subordinates (Don). Ted wants to convey to Don that his performance is no longer up to standard, but that he (Ted) genuinely wishes to give him a second chance. While the fact that one feedback process (that between Ted and Don) is embedded in a second (that between the respondent and Ted) adds to the complexity of the exercise, it has the advantage of making it possible to compare the standards that respondents use to judge Ted, with those that guide their own feedback practices. In this way a comparison of respondents' espoused theories and theories-in-use is possible.

Participants learn about the way Ted conducted the interview via a memo which provides an accurate and representative sample of what he had said to Don. The memo and the instructions to respondents are reproduced below.

The Ted–Don Exercise

The following is a transcript of some of Ted's statements to Don. Please assume that they are a valid sample of the statements that Ted made.

Don, your performance is not up to standard, [and moreover] you seem to be carrying a chip on your shoulder.

It appears to me that this has affected your performance in a number of ways. I have heard words like lethargy, uncommitted and disinterested used by others in describing your recent performance. Our seniors cannot have those characteristics.

Let's discuss your feelings about your performance.

Don, I know you want to talk about the injustices that you believe have been perpetrated on you in the past. The problem is that I am not familiar with the specifics of these problems. I do not want to spend a lot of time discussing something that happened several years ago. Nothing constructive will come of it. It's behind us.

I want to talk about you today, and about your future in our system.

In completing the exercise, respondents are asked to imagine that Ted comes to them and asks, "How well do you think I dealt with Don?" In answering, they are to assume that Ted wants to learn. Respondents write their answer to Ted's question on sheets of paper divided into two columns. On the right-hand side, they write exactly what they would say to Ted and what they would expect him to reply. On the left-hand column, they record any thoughts or feelings which they do not express to Ted.

The exercise captures the multifaceted character of many problems of practice, for it presents a set of interrelated problems and attempted solutions, rather than one discrete problem. The first problem is that of Don's poor performance, which Ted has attempted to solve through the feedback reported in his memo. His attempted solution becomes a second problem, however, because the vast majority of respondents believe that his feedback is incompetent (Argyris, 1982, pp. 42–44). Although it is this second problem that respondents are asked to address, they cannot do so in isolation from the first, because their feedback to Ted will reflect their theories about Don's performance. The more their explanations for Don's performance differ from those of Ted, the more negative their feedback to the latter is likely to be. The correlation between their explanatory theory and their theory of action for helping Ted is likely to be far from perfect, however, because the latter will also reflect their beliefs about how negative feedback should be delivered.

Both problems in this exercise are ill-structured in the ways previously described. There is uncertainty about what to say to Ted because there are dozens of possible explanations of Don's poor performance and as many plausible theories about how to help Ted (assuming that his performance is judged to be problematic). Ted and Don's problems, in common with most client problems, "are rich enough to provide some support for all but the most outrageous hypotheses" (Arkes, 1981). The available information does rule out some possible interpretations, however. For example, the memo tells us that Ted did not listen to aspects of Don's story, so it would be wrong to argue that Ted was ineffective because he spent too much time listening. On the other hand, the statement "Your performance is not up to standard" could be treated as evidence for either Ted's effectiveness or ineffectiveness, depending on the background theory of the analyst.

The Ted–Don exercise also illustrates how there is ambiguity about the nature and amount of information needed to solve ill-structured problems. The memo from Ted tells us what Ted said but not how Don reacted. Analysts who believe that interpersonal effectiveness can best be judged from its consequential effects will find this gap a substantial barrier to problem-solving. Those who bring process theories to bear, are more likely to see the data as sufficient. There is, therefore, not only uncertainty about how to interpret the available information, but uncertainty about what additional information, if any, is required.

Carol's Solution to the Problem

The data reported on page 33 represent the way one school principal (Carol) responded to this exercise. Carol had been principal of a large urban high school for two years when she completed this exercise as part of her involvement in the case study reported in Chapters 4 and 6.

Carol solved the problem of how to give Ted feedback in about twenty minutes. In the left-hand column we see some of her reasoning about Don, about Ted's effectiveness, and about how to frame her feedback. The right-hand column describes the conversation she would have with Ted.

Carol solved the problem of what to say to Ted by developing an explanation of Don's poor performance, which she then used as a standard by which to judge the effectiveness of Ted. Two different explanations of Don's poor performance are entertained; passing consideration is given to the possibility that Don may be in bad health (l. 39), and a much richer set of hypotheses are developed about how Don may be desperate to talk about the past (l. 40), need someone to listen to him (l. 41), and not be able to feel part of the team until someone does so (l. 79).

Carol then uses her theory about Don to evaluate Ted's feedback skills. If Don's poor performance is attributable, at least in part, to others' failure to listen to his grievances, then effective help requires that this be remedied. Carol concludes that Ted has not listened, and explains this failure with reference to his lack of preparation (l. 86), his prejudgement of Don (ll. 13 and 79), his belief that the past is unimportant (l. 36), and the attitudes and skills which he brings to the interview.

Carol's Solution as a Theory of Action

In analysing Carol's solutions as a theory of action, an important question arises about the nature of the evidence which the exercise provides. Espoused theories of action are derived from statements of intent, or narrative accounts that do not report actual speech. For example, Carol's wish not to do the same thing to Ted as he did to Don (l. 19) is a statement of intent that would contribute to the formulation of her espoused theory of feedback. Theories-in-use, in contrast, are derived from evidence about how people actually behave as captured through observation or recording. The exercise which Carol completed

Carol's Conversation with Ted

	Unexpressed thoughts	*Conversation*
		Carol: Ted, you've asked me to comment on your conversation with Don. . . .
5		*Ted:* Yes. I want to know if what I said was effective.
10	Ted will probably be very defensive. I feel that he thinks that he has done a good job.	*Carol:* I hope you're prepared for me to be critical because I don't think you handled Don's situation well?
15	I think he's already made judgements on Don's ability and potential and acted on them.	*Ted:* Why not! I got to the point didn't I?
20	I guess I'm getting a bit angry with Ted. Better not do the same thing he did with Don!	*Carol:* Yes, I guess you did—but I think you took charge of the interview very firmly and *told* Don a lot of things. . . . Did you think it was necessary to ask Don about what he thought about the critical things you were saying about his attitude?
25	I think Ted should be substantiating any criticism he made with evidence and listening to what Don's perceptions of problems are.	*Ted:* Not really. I have a pretty good idea of what he's not doing. For one, he's not pulling his weight and I've been told that by a lot of people.
30	Did Ted even bother to find out details of past problems before he talked to Don?	*Carol:* Right. Can we get back to that point later? Tell me about this "chip on the shoulder" you mentioned.
35	I wonder if he accords it any importance?	*Ted:* Well apparently, something happened a long way back . . . well before my time. . . .
	Does Don have a health problem?	
40	It's possible that Don is desperate for someone to finally listen to his side of the story. I think he wants to get this across to Ted but Ted's not willing to listen.	*Carol:* Perhaps Don would have felt you were listening to his point of view (I hope you agree that he has one) if you had let him continue at that point.
45	I think Ted is treating this as a "task" and not considering that a *person* is involved.	*Ted:* Yes, perhaps, but I didn't want to waste time rehashing old grievances. Besides, the main idea was to get him to think about the future.
50	I wonder if Ted is scared to get in too deep. Is the "chip" something he can't	*Carol:* I gather, however, that he did want to talk about the past. Now you did at one stage suggest that he talk about his *feelings* about his performance?

Carol's Conversation with Ted—continued

	Unexpressed thoughts	*Conversation*
55	do much about anyway?	
		Ted: Yes I did, but he immediately began to blame others and brought up this incident which happened years ago.
60	At least he could jolly well listen! Even if he thinks or knows the problem is Don's . . ., he's supposed to be helping solve it.	*Carol:* You didn't hear what he had to say then?
		Ted: No.
65		*Carol:* Do you think Don feels any different after talking to you?
70	Perhaps Ted is feeling a bit guilty – I may be able to persuade him to try again – for what it's worth!	*Ted:* No, probably not. But I had to spell out what I thought was important to everybody *right now.*
		Carol: Which is?
75		*Ted:* To get him to see sense. He's got to begin behaving like a senior staff member.
80	I really feel that Ted thinks Don's a hopeless case but doesn't want to be seen as "not trying"!	*Carol:* And do you think he feels like one of the team?
85		*Ted:* Who knows? Probably not. I guess he still feels the same way about the company and the other senior staff—but he does know his job is on the line.
90	I think Ted has realised that he has not done his homework and has not really achieved what he set out to achieve.	*Carol:* Well, Ted. If that's what you set out to convey to him, fine. You were very effective. But I thought you were genuinely hoping to give him a second chance.
	I think he's feeling concerned that he's not really supported Don and this was what was intended in the first place.	
95		*Ted:* Well, perhaps I didn't get that across very well. Right! Should I try again?
	Is the company really using Ted to get rid of Don?	
100		*Carol:* Well, if you really want to, you'd hear what he has to say and perhaps try to use it constructively to solve his problem.
105	I think Ted has got to try again and I feel he will make a genuine effort to hear what Don has to say.	*Ted:* Will do! I'm prepared to do that much as least.
	As to solving the problem, who knows!	

did require her to produce a conversation, but in simulated rather than in *in vivo* conditions. The question arises, therefore, whether analysis of the conversation yields a theory-in-use or an espoused theory of action. The former position will be taken because the right-hand side yields actual speech rather than reports of speech, and as we shall see, can be evaluated against Carol's own espoused theory of feedback. The question remains, however, as to whether this theory-in-use accurately predicts Carol's behaviour in real negative feedback situations. This issue is addressed in Chapter 4 as part of the analysis of Carol's leadership style. In general, however, Argyris's research shows that inferences made from this exercise are generalisable to real feedback situations (Argyris, 1982, pp. 50–60).

The left-hand column of the transcript gives us important clues to Carol's espoused theory of negative feedback, because it includes some of the injunctions she gave herself about how to communicate to Ted. She wants to persuade him to try again (l. 70), without making him defensive. In achieving these goals, she must avoid using the same strategies that Ted used on Don (l. 19).

Carol's theory in use is derived by inferring the constraints that she actually set on the solution, the strategies she employed in attempting to satisfy those constraints, and the associated consequences, both intended and unintended. Carol began the conversation with Ted by inoculating him against the criticism that was to come (l. 9). She then conveys her criticism forthrightly, and asks a series of questions, some of which presuppose the truth of her own views, and some of which appear to be genuine inquiries (l. 67). She maintains control of the agenda by setting aside Ted's concerns, and telling him what he should think. Finally, her thoughts on the left-hand side reveal a pattern of avoidance of discussion of her own feelings and those she attributes to Ted. It seems plausible to claim that these strategies serve a goal of winning Ted over to one's point of view, while avoiding discussion of emotional issues.

To complete the model of Carol's theory, we need to establish the consequences of giving negative feedback via a strategy that involves telling another what to do, while simultaneously ignoring one's own and the other person's intellectual and emotional concerns about the wisdom of that advice. The transcript provides some relevant information about consequences, and some of the gaps can be tentatively filled by drawing on relevant background theory. It is important to remember, however, that consequences such as Ted's reactions will be the result of an interaction between Carol's theory of giving and Ted's theory of receiving negative feedback. As such, they cannot be directly attributed to Carol's theory-in-use. The most obvious consequence of the interaction is that Carol has been successful in (apparently) winning Ted over to her point of view. But there is considerable uncertainty, as Carol herself acknowledges, about whether the problem has been solved (l. 107). Her uncertainty seems realistic, given that Ted held views that were contrary to Carol's and the differences were neither acknowledged nor adjudicated. Ted may have agreed to do something which he does not fully understand and to which he is not committed. In either case, he may have difficulty implementing the advice in the way Carol would wish. The theory-in-use that Carol employed prevented her from discussing either the wisdom of her advice or Ted's ability to implement it. If errors occur at either level, Carol's theory has severely reduced the chance of resolving the problem of Don's poor performance.

Given the literature on the difficulties associated with negative feedback, it is also relevant to consider the consequences of Carol's theory-in-use for the relationship between her and Ted. The transcript provides no directly relevant evidence, but again we can employ relevant background theory to develop hypotheses. If Ted acts dependently towards those from whom he seeks feedback, Carol's strategy will not damage their relationship. If not, he may feel that he has been talked into seeing the error of his ways without the opportunity to debate his objections to Carol's advice. In this case he may feel prejudged, and reluctant to seek Carol's help in the future.

Theory Appraisal: How Good is the Solution?

The emphasis so far, on the ill-structured and uncertain nature of most educational problems, may suggest the inevitability of some type of relativist stance towards theory appraisal. If the evidence relevant to the problem is, as Arkes (1981) claims, sufficiently rich to support all but the most outrageous hypotheses, how on earth can practitioners and researchers choose between them?

One answer is that they choose on a range of non-rational grounds, such as ideology, familiarity or precedent. Another is that they do not choose at all, but are constrained to act in conformity with particular theoretical traditions, through the power of professional or political regulation. While there is considerable descriptive merit in these arguments, they are inadequate for a methodology which seeks to explain why educational problems arise, and to contribute in a non-impositional manner to their resolution. Many such problems arise precisely because people operate from competing theories in situations where co-ordinated action is needed, or because one theory has come to dominate in a way that prevents examination of its adequacy. If researchers are to provide a better theory, rather than just privilege one of the contenders in the problem situation, they need to acknowledge the possibility of theory adjudication and propose a method for accomplishing it.

The remainder of this chapter presents a set of criteria for adjudicating theories intended to serve as solutions for practical problems. The criteria are designed to suit the particular purpose of PBM, which is to understand problems in ways that they can be resolved. They are based on the work of Argyris and Schön (1974, pp. 20–35) and, to a lesser extent, on that of the philosopher of science William Lycan (1988). My purpose is not to give a lengthy philosophical defence of each criterion, but to describe each one prior to illustrating its application to the evaluation of Carol's solution to the problem of how to give feedback to Ted. At times, the contrasting theories held by Ted are also used to explain the operation and limitations of the criteria.

Explanatory Accuracy

On this criterion, one theory is more adequate than another if it can provide a more accurate causal account of the phenomenon it seeks to explain. Such an account shows how particular events and processes combine under particular conditions to produce the phenomenon of interest.

Carol believes that Don's poor performance is caused by a lack of listening; *ipso facto*, his performance should improve if relevant actors hear his side of the story. Ted, on the other hand, believes that it is precisely this dwelling on the past that is causing Don's poor performance. In theory, we should be able to apply the criterion of explanatory accuracy to adjudicate between these two views. In practice, however, both these explanations are consistent with the meagre available evidence. Don is preoccupied about the past, and at least some people have been unwilling to listen to his preoccupations. We also have relevant background theories to support either position; a version of reinforcement theory would support Ted's view that attention to nonproductive behaviour would strengthen this tendency. Maslowian theory, on the other hand, would suggest that attention to Ted's grievances would satisfy his needs for recognition and esteem, so that he could then direct energy towards higher level achievement needs (Maslow, 1968).

The impasse could be broken by obtaining evidence about Don's behaviour under the conditions suggested by each theory. If people have listened to Don's grievances in the past, with no consequent improvement in his performance, we have reason to judge Ted's theory as superior to Carol's on this criterion of explanatory accuracy. It is equally possible, however, that such evidence does not resolve the impasse, because Ted and Carol cannot agree about whether these past attempts really count as "listening to Don". Carol may then conduct an "experiment", and demonstrate what she means by "listening to Don". If Don still does not improve, she could still legitimately save the theory by arguing that some other condition had not been satisfied; maybe Don needed more time spent with him, or the improvement in his performance would not be evident until a later date. While these moves to save a theory are logically defensible, the particular circumstances of each case determine whether or not they represent reasonable scepticism or defensive manoeuvres on the part of the advocate.

In conclusion, the criterion of explanatory accuracy has done some work in this example, such as suggesting what additional data would facilitate adjudication, but in many cases like this one the criterion will not yield a definitive judgement about the better theory. Let us see then, whether the remaining criteria do more work in this case.

Effectiveness

A theory of action is effective if it produces the intended consequences without violating important constraints. This criterion is, therefore, broader than a criterion of goal attainment, because it also tests whether the goal has been achieved at an acceptable cost.

When a theory of action has a causal theory of the problem embedded within it, there is considerable overlap between this criterion and that of explanatory accuracy. If the causal theory is wrong, and the theory of action presupposes its validity, then the theory of action will be ineffective. There are many theories of action, however, which operate independently of a causal theory of the problem itself. For example, we can construct theories of action to help Don, which are effective even though we have no knowledge of the cause of his poor performance. In addition, we may know the cause (he has cancer), but construct an effective theory of help that makes no use of that knowledge. Practitioners frequently face situations in which they either do not know the cause of the problem, or they know it, but are unable to employ that knowledge in their attempts to resolve the problem. We need a criterion of effectiveness as well as of explanatory accuracy, because success on the first criterion is neither a sufficient nor a necessary condition for success on the latter.

When asking about the effectiveness of a theory of action we are essentially involved in tracing the consequences of the theory and asking whether they satisfy the constraints that were set. Argyris and Schön (1974) describe this as follows:

> A theory-in-use is effective when action according to the theory tends to achieve its governing variables. Accordingly, effectiveness depends on: the governing variables held within the theory; the

> appropriateness of the strategies advanced within the theory, and the accuracy and adequacy of the assumptions of the theory (p. 24).

The consequences of Carol's theory-in-use were discussed in the previous section, so they will be only briefly summarised here. While Carol achieved her goal of winning Ted over to her point of view, she did so at the cost of violating some of the other constraints on the problem solution. First, although we lack evidence about Ted's reactions, it is likely that her judgemental approach was incompatible with her desire to avoid upsetting him. Second, her persuasive style was incompatible with the conditions needed to help Ted learn how to assist Don more effectively. This analysis shows how practitioners can unwittingly sustain the very problems they are seeking to solve. Carol's theory of action maintains rather than resolves the problem of Ted's poor performance, and indirectly that of Don, because it does not provide the conditions required for him to effectively implement her advice, or to test whether the advice she is giving is sound. Evidence from hundreds of responses to the Ted–Don exercise (Argyris, 1982, pp. 42–44) and from real supervision and appraisal situations (Bridges, 1986) suggests that feedback senders typically attempt to persuade receivers of the correctness of their views through a range of nice or nasty strategies. This type of theory of action is ineffective in that it has a high probability of either not improving performance, damaging the relationship between feedback sender and receiver, or both.

Coherence

The third criterion of theory adequacy examines the coherence of a theory of the problem with our theories about all other problems. According to Lycan (1988, p. 130), we should prefer one theory to another if "it squares better with what you already have reason to believe". Note that Lycan talks about consistency with what we already have reason to believe; that is with theories that have already passed a relevant test of epistemic virtue, not about consistency with what we already believe. There is no value in consistency when one's existing belief system is faulty. Seeking internal consistency involves eliminating inconsistency between aspects of one's global theory; seeking external consistency involves eliminating inconsistency between theory and evidence. This coherence criterion is the same as that presented in Chapter 1 for making judgements about the improvement of practice. Following Walker (1987), improvement was said to have occurred when we solve problems in ways that are consistent with solutions to our whole problem set; in other words, when the way we solve one problem does not make it harder to solve others.

TABLE 2.1. *The Congruence of Carol's Theory of Action*

	Espoused theory	Theory-in-use
1.	Treat Don as a person and explore his feelings.	Treat Ted as a task and ignore his feelings.
2.	Convey opinions with examples, give reasons and check Don's reactions.	Ask some genuine questions of Ted. AND Assert one's opinions without gaining Ted's reactions.
3.	Listen to Don's side of the story.	Pursue one's own theory without listening to that of Ted.

It is much easier to detect incoherence than to decide how to remedy it, because incoherence can be eliminated by altering either or both of the inconsistent elements. Take the special case of incoherence between espoused theory and theory-in-use, which Argyris calls incongruence. Table 2.1 summarises some of the incongruence between Carol's beliefs about how feedback should be given (inferred from the advice she gives to Ted) and the way she actually delivers it. In advising Ted, she espouses treating him as a person with feelings, listening to him, and communicating the basis of her opinions. The way she herself gives this feedback, however, violates her own advice. Carol can remedy the incongruence between her espoused theory and theory-in-use by altering either or both of these two elements. If Carol alters her espoused theory, congruence is brought at the price of ineffectiveness. As Argyris and Schön (1974) explain, congruence is only a virtue when the element involved itself meets the other criteria of theoretical adequacy.

> There is no particular virtue in congruence alone. An espoused theory that is congruent with an otherwise inadequate theory-in-use is less valuable than an adequate espoused theory that is incongruent with the inadequate theory-in-use, because then the incongruence can be discovered and provide a stimulus for change (pp. 23–24).

It is possible, therefore, to increase coherence in ways that violate other criteria for theory appraisal, such as explanatory accuracy or effectiveness. Since this criterion operates over the whole of our theoretical knowledge, however, and not over isolated sectors of that knowledge, it would soon begin to bite. If Carol reduced incongruence between her espoused theory and theory-in-use by altering the former, the revised theory is likely to be incoherent with other relevant theory about, for example, how people learn from others' advice. The point can be made in exaggerated form by considering the theories of schizophrenics; their paranoid beliefs may be self-consistent and congruent with their behaviour. Their theories are still incoherent, however, because they violate our best theories about how the world works, and as a result, schizophrenics are severely restricted in their capacity for independent living.

The coherence criterion demands theories be appraised by considering knowledge that lies beyond the framework of the theory itself. Under effectiveness we examined the way a theory served the constraints that were set in order to solve the problem. The coherence criterion, in principle, sets all our best theory as a constraint on theory formulation, and thus is a force for knowledge integration, and for questioning the parameters that comprise our current formulation of particular problems.

Improvability

If we know that many educational problems are ill-structured, then our theories of such problems need to cohere with this knowledge, while at the same time reducing this uncertainty so that the problem may be solved. Given a shortage of information, ambiguity of evidence and numerous possible constraint structures, there is a high probability of error in our theories. There is merit for the growth of knowledge and for problem-solving, therefore, in the development of theories which are open rather than closed to the possibility of error and revision. The criterion of improvability assesses this quality. A theory of action illustrates this quality to the extent that it simultaneously suggests what to do, and how to detect inadequacies in its own advice.

Argyris's writing provides some specific guidelines about how to assess the improvability of a theory of action. The table below provides a version of these guidelines, and the headings "Theory openness" and "Theory closedness" are shorthand descriptions for the qualities of theories which promote and inhibit improvement. The strategies on the left-hand side of Table 2.2 encourage feedback from the human and non-human environment about the adequacy of the theory, while those on the right restrict it. A distinction must be drawn, however, between two different types of feedback, for they result in different types of learning. The constraint structure of a theory of action establishes the data that are critical for determining whether or not the theory is effective. Carol, for example, monitors Ted's reactions so that she can learn whether he is becoming defensive and whether he has been persuaded to try again. She adjusts her strategies accordingly, in order to remain on target, but the definition of what it is to be on target

TABLE 2.2. *The Improvability Criterion of Theory Appraisal*

	Theory openness	Theory closedness
1.	The claims made by the theory are publicly available for examination.	The claims made by the theory are kept private or made so vaguely and tentatively they are hard to know.
2.	The examples, evidence and reasoning processes which support claims are made public so the adequacy of the claims can be checked.	The reasoning and evidence that support claims are not available. Inferential claims are treated as "self-evident".
3.	Evidence or argument which could potentially disconfirm the theory is sought and publicly evaluated.	Potentially disconfirming evidence or argument is ignored or discouraged.

remains outside the critical feedback loop. Argyris describes this type of theory improvement as single-loop learning, because it is restricted to monitoring how effectively strategies satisfy a given constraint structure. Double-loop learning involves changes to the values, goals and key assumptions that make up that constraint structure. Carol will not be able to resolve the Ted–Don problem without engaging in double-loop learning because her theory-in-use incorporates an incompatible constraint structure. She cannot simultaneously persuade Don to her point of view, avoid upsetting him, and help him to be more effective in his approach to Ted.

Single-loop learning is highly efficient, since it allows for a theory to be revised while preserving its central tenets. Double-loop learning is highly disruptive, since it destroys the routine the practitioner had established for understanding and resolving the problem:

> Double-loop learning does not supersede [*sic*] single-loop learning. Single-loop learning enables us to avoid continuing investment in the highly predictable activities that make up the bulk of our lives; but the theory builder becomes a prisoner of his programs if he allows them to continue unexamined indefinitely. Double-loop learning changes the governing variables (the "settings") of one's programs and causes ripples of change to fan out over one's whole system of theories-in-use (Argyris & Schön, 1974, p. 19).

The application of the criterion of improvability to Carol's theory-in-use is summarised in Table 2.3. The left-hand side lists the major claims made about the cause of Don's poor performance and how to help him, and on the right-hand side the learning (or non-learning)

TABLE 2.3. *The Criterion of Improvability Applied to Carol's Theory-in-Use*

	Theoretical claims	Learning strategies
	Claims about Don	
1.	Don may have a health problem.	Not disclosed (l. 38)
2.	Don may be desperate for someone to listen.	Partially disclosed. Question presupposes desirability of listening (l. 40).
3.	Don wanted to talk about his past.	Disclosed. Illustrated via Ted's memo (l. 50).
4.	Don needs to feel like one of the team.	Disclosed. Question presupposes validity of claim (l. 79).
	Claims about how to help Don	
5.	Ted has prejudged Don.	Not disclosed (l. 13).
6.	Ted took charge of the interview and told Don things.	Disclosed, not illustrated or checked (l. 16).
7.	Ted did not seek feedback from Don, and this is a problem.	Disclosed through question which treats claim as self-evident (l. 19).
8.	Ted did not listen to Don's perceptions of his problems and listening is important.	Loaded question treats claim as self-evident (l. 40).
9.	Ted did not do his homework, and this jeopardised his effectiveness.	Not disclosed (l. 30).
10.	Ted is task oriented rather than person oriented, and this reduces his effectiveness.	Not disclosed. Related questions assume validity of claim (l. 47).

strategies associated with each claim. Claims 1–4 concern Don's poor performance, and claims 5–10 are the first five of Carol's many claims about Ted. The table shows that the great percentage of the theoretical claims that made up Carol's theory-in-use were accompanied by non-learning or single-loop learning strategies. For example, Carol checks her claim about listening by checking her perception that Ted has not done so, while simultaneously presupposing the validity of her assumption that listening is important (claim no. 8, Table 2.3).

One could argue that the improvability criterion becomes less important the more the subject specific content of a theory meets high standards of explanatory accuracy, effectiveness and coherence. If a theory is good enough, learning features of the theory will be redundant. The problem with this argument is that we do not know for certain when our theories have this quality, and by presuming that they do, we are blind to the situations that would otherwise trigger their revision. The contrary view is also possible, viz. that the criteria of explanatory accuracy, effectiveness and coherence are redundant, because a theory that meets the learning criterion will, over time, come to possess these other qualities, through the process of error detection and correction. To argue this line seems to deny the systematic growth of knowledge and to portray theorists as starting afresh each time. It is highly inefficient, if not unethical, to practise (science or education) by ignoring prior knowledge about the problem. Bad theories about the world, in education and the social sciences at least, tend to have at least as much force as good ones, because of their powerful self-fulfilling properties. The improvability criterion of theory appraisal is an insurance on the rest, not a substitute.

Summary

The major concern of this chapter has been how researchers can understand educational problems in ways that are likely to promote their resolution. It was argued that problematic practices and outcomes should be viewed as the results, intended or unintended, of practitioners' prior problem-solving attempts. If we understand the problem for which the problematic practice was an intended solution, we can understand what it is that generates and sustains the practices we wish to alter.

The argument was developed by showing how most educational problems are ill-structured and how they are solved by generating a constraint structure which defines the parameters of acceptable solutions. Such constraint structures and their associated solution strategies constitute the practitioner's theory of action for the problem. Successful investigation of the links between the problem and the relevant theory of action requires a careful distinction between espoused theory and theory-in-use. When espoused theories are incongruent with theories-in-use, they contribute little to the explanation of the problem. Accurate reconstructions of theories-in-use provide researchers with access to the knowledge that practitioners themselves use in the situation. They explain practitioners' behaviour by revealing the constraints they attempted to satisfy, and the consequences, both intended and unintended, of the choices they made.

Research which contributes to the resolution of educational problems requires a methodology for the evaluation as well as the reconstruction of theories of action. The second half of the chapter outlined and illustrated four criteria for this task. The first criterion asks whether the theory of action incorporates an adequate explanatory theory for the problem it seeks to resolve. The second asks about the effectiveness of the theory of action, where effectiveness is defined by the constraint structure of the problem. The third coherence criterion alerts us to the possibility that the theory may violate constraints that lie outside the constraint structure of the original problem. The final improvability criterion was included in the set, in response to the uncertain and ill-structured nature of educational problems, and the consequent adaptive value of theories that provide for their own revision. In short, theories of action are judged inadequate in PBM if they incorporate faulty understandings of the nature of the problem, ineffective solution strategies, or if in resolving one problem they exacerbate or create others. Any such difficulties are compounded if the theory incorporates features that inhibit inquiry into and revision of its own tenets.

An obvious limitation of the argument of this chapter is that educational problems do not typically arise as the result of the theory of one or two practitioners. While there were heuristic advantages in using a relatively simple example to introduce the argument, it needs to be translated to an educational world of interacting groups, organisational routines and cultures, and externally constrained choice. How are theories of action linked to educational problems in this context? The question is central to the two school-based case studies which are presented in Chapters 4, 5 and 7. The immediate concern of the next chapter, however, is to further develop and defend problem-based methodology.

References

Argyris, C. (1982). *Reasoning, learning and action.* San Francisco: Jossey-Bass.

Argyris, C. & Schön, D. (1974). *Theory in practice: Increasing professional effectiveness.* San Francisco: Jossey-Bass.

Argyris, C., Putnam, R. & McLain Smith, D. (1985). *Action science.* San Francisco: Jossey-Bass.

Arkes, H. R. (1981). Impediments to accurate clinical judgement and possible ways to minimize their impact. *Journal of Consulting and Clinical Psychology*, **49** (3), 323–330.

Bridges, E. M. (1986). *The incompetent teacher* (Stanford Series on Education and Public Policy). Lewes: Falmer Press.

Chi, M. T., Feltovich, P. J. & Glaser, R. (1981). Categorization and representation of physics problems by experts and novices. *Cognitive Science*, **5**, 121–152.

Cusella, L. P. (1987). Feedback, motivation and performance. In F. Jablin, L. Putman, K. Roberts and L. Porter (Eds.), *Handbook of organizational communication: An interdisciplinary perspective.* California: Sage Publications.

Frederiksen, N. (1984). The real test bias: Influences of testing on teaching and learning. *American Psychologist*, **39**, 193–202.

Lycan, W. (1988). *Judgement and justification*. Cambridge: Cambridge University Press.
Maslow, A. (1968). *Toward a psychology of being*, 2nd ed. Princeton, NJ: Van Nostrand.
Nickles, T. (1981). What is a problem that we might solve it? *Synthese*, **47** (1), 85–118.
Pondy, L. & Huff, A. S. (1985). Achieving routine in organizational change. *Journal of Management*, **11**, 103–116.
Robinson, V. M. J. & Halliday, J. (1988). The relationship of counsellor reasoning and data collection to problem analysis quality. *British Journal of Guidance and Counselling*, **16**, 50–62.
Simon, H. (1973). The structure of ill-structured problems. *Artificial Intelligence*, **4**, 181–201.
Voss, J. F. & Post, T. A. (1988). On the solving of ill-structured problems. In M. T. Chi, R. Glaser and M. J. Farr (Eds.), *The nature of expertise*, pp. 261–285. Hillsdale, NJ: Lawrence Erlbaum.
Walker, J. C. (1987). Democracy and pragmatism in curriculum development. *Educational Philosophy and Theory*, **19** (2), 1–10.

3

Defending and Developing Problem-based Methodology

Problem-based methodology can be most usefully elaborated at this point by raising and replying to a number of possible objections. The first half of the chapter defends PBM against the charges that it is concerned with technical problem-solving, that it neglects the wider social context, and that it formulates educational problems on too small a scale. The second half of the chapter completes the description of the conceptual framework of PBM, by describing how critical dialogue governs its social relations of inquiry.

THREE POSSIBLE OBJECTIONS

Mere Problem-solving?

The title "mere problem-solving" captures the pejorative evaluation that some social scientists bestow on forms of inquiry directed towards the resolution of problems. Defending problem-based methodology against such criticism will involve describing how such critics understand problem-solving and how that understanding and its associated criticisms do not apply to the notion of problem-solving which is central to this book.

The criticisms are associated with the view that problem-solving is a search for the most efficient and effective means to given ends. This view of problem-solving as an exclusively technical or instrumental activity is flawed because it rules out critique of the normative and assumptive framework within which a problem is understood. For example, it may be impossible to find a solution to the problem of truancy from high school, while it is assumed that any such solution must be directed towards increasing student attendance at existing class programmes.

The influence of this narrow view of problem-solving has been greatly increased through its inclusion in the third level Open University education course "Policy Making in Education" (Open University, 1986). In this text, Cox (1981, pp. 128–129) explains that problem-solving involves making institutions work smoothly by dealing effectively with particular sources of trouble that arise within them. The prevailing power and social relations of such institutions are treated as a stable backdrop against which discrete problem-solving activities can take place. Restated in terms of a constraint inclusion account of problems, Cox is saying that problem-solving is restricted to a single-loop search for a solution that fits the existing constraint structure of the problem.

The constraint inclusion account of problem-solving in PBM is qualitatively different from that described by Cox, because it self-consciously incorporates an evaluation of the existing and of alternative constraint structures. Where the problem cannot be solved under the existing structure, a new theory of the problem is formulated and evaluated according to the criteria of explanatory accuracy, effectiveness, coherence and improvability, as outlined in Chapter 2.

Theories of Action and the Wider Context

Most educational research formulates educational problems in ways that have nothing to do with the theories of action of those involved. For example, correlational studies of tracking tell us the characteristics of those who are placed in particular teaching groups, and structurally-oriented critical research reveals the reproductionist consequences of such practices. Neither type of research will focus on the theories of action that lead those responsible to organise teaching and learning in this way. What are the arguments for doing so?

First, as I have already argued in Chapter 1, theories of action may be causally implicated in the problem, and if this is the case, the problem cannot be resolved, unless coercively, while those theories are in place. Second, this approach identifies particular agents who are, or can become, responsible for the problem situation, and for working with the researcher to resolve it. Intervention processes cannot operate from highly abstract depersonalised accounts about how, for example, the capitalist economy or institutional racism maintains achievement-based tracking arrangements. This is not to say that such factors may not be important, but if they are, they need to be incorporated within an actionable analysis that shows how they are mediated by particular agents in the particular problem situation, and how they are alterable by those agents. One of the reasons, I believe, that much educational research fails to deliver on its practical promise, is that its formulations do not identify actionable variables, or the agents, whose concern, commitment and ability to learn can drive the problem resolution process.

The third reason for formulating problems in terms of theories of action is that this approach is compatible with a process of change that is educative rather than coercive. People act for reasons, and if those actions are implicated in the problem, they will not act otherwise until they are satisfied with the validity of the researcher's critique and the adequacy of the suggested alternative.

There are, of course, arguments against formulating problems around theories of action, as well as arguments in favour. One of the more important objections to this approach is that it focuses too much on individual understanding and action, and ignores or underestimates the importance of context and social structures in the development and maintenance of educational problems. The criticism can be developed in the context of the Ted–Don case by imagining it is set in a hierarchically organised secondary school, in which Ted is an autocratic principal and Don a relatively inexperienced Head of Department. A structurally oriented sociologist might argue that our analysis of the case should not be confined to the interaction between Ted and Don, but should include the patterns of institutional authority and power which foster Ted's unilateral and punitive approach. Such researchers might argue that the problem is not just Ted's poor leadership style, but a hierarchical school organisation which expects such "strong" leadership.

The first point to note about this objection is that it is not about the relative merits of theory of action and structural explanations of Ted's problem, but about the adequacy of focusing only on the former. A theory of action explanation of Ted's behaviour links the problematic practice (the way he gave feedback to Don) to the understandings and values that Ted brings to the situation. These might include such beliefs as "effective leadership is strong and uncompromising". The objection is that we should know why Ted holds such beliefs and values, as well as know what they are (Fay, 1975, pp. 83–91). Theory of action explanations, in short, are alleged to be insufficient, because of their omission of the wider social structure.

Since any explanation can be deepened by asking how it itself can be further explained, we need to understand the purpose for which theory of action explanations are alleged to be insufficient. One could argue, for example, that analyses which show how theories of action are linked to organisational cultures make a greater contribution to social theorising than those that do not. This would not be a telling criticism of problem-based methodology, however, for its primary purpose is the analysis and resolution of practical problems, and not the development of social theory. (This is not to say that PBM would not contribute to social theory, only that that is not its primary purpose. More is said on this in a discussion of the generalisability of theory of action explanations in Chapter 6.) The criticism would be more telling if it could be shown that these practical purposes were jeopardised by omission of the wider context. Do theory of action explanations fail to meet the criteria previously defended as essential to these purposes? In the context of the Ted–Don dialogue, this question asks whether an account of Ted's theory-in-use provides an adequate explanation of the way he gave feedback to Don. If we cannot alter Ted's theory-in-use without also altering the social context in which he is located, then explanations which omit the latter are insufficient. Ted may find it impossible to alter his leadership style in a school where strong tough leadership is expected, rewarded and constantly modelled. On the other hand, the linkages between Ted's position and the rest of the school hierarchy may be sufficiently loose to allow Ted to experiment with alternative ways of interacting with his staff, in contradiction to the prevalent culture. Given new skills and favourable evidence about their consequences, Ted may eventually recruit others to a counter-culture which challenges the prevailing views about what it is to be an effective leader.

In summary, my reply to the objection that PBM neglects the wider context has been that its purposes are not necessarily jeopardised, and are frequently enhanced, by theorising which centres on the theories of relevant agents and which does not necessarily seek to broaden the analysis beyond this point. However, theory of action explanations may turn out to be insufficient when other factors prevent those involved from making the necessary changes to their reasoning and action, or when those changes do not result in the predicted problem improvement. Incompleteness is relative to a particular purpose and context, and there is no *a priori* reason why structural factors must supplement theory of action explanations.

My second reply to the objection about the neglect of structural factors is based on the following important distinction. To focus on individuals, as in the formulation of theories of action, is not necessarily to individualise one's problem analyses, in the sense of being insensitive to the way in which social context impinges on individual thought and action. The claim that theory of action explanations neglect structural factors, rests on an overly sharp distinction between structural and interpersonal factors. If Ted's action is tightly determined by the wider context, then those influences should be revealed through a careful analysis of his understandings and actions in the problem context. If Ted is aware of these connections, there will be cues to relevant structural influences in the way he describes and defends his own practice. He might claim, for example, that he is expected to "act tough", or that "that is the way we do things around here". Even if Ted is unaware of these influences, researchers could still detect them through careful observation of his behaviour and of the antecedents and consequences of the strategies that he employs. While theory of action explanations stick closely to the immediate context of the problem situation, this does not imply that they ignore structure, because structure operates through the actions and understandings of individuals.

The Scale of the Problem

Theory of action formulations of educational problems may be unacceptable to some critics, not because they omit structural variables, but because the focus on particular agents produces a small scale rather than a more holistic analysis of the problem. How can analyses of the interactions and reasoning processes of a few individuals, even when appropriately linked to structural influences, tell us anything significant about educational problems which exist on a massive scale?

In this section, this criticism will be addressed through discussion of the implications of the size of the problem for the probability that the practice-oriented researcher will contribute to the improvement of practice. I argue, following Karl Weick (1984), Lindblom (1979) and Lindblom and Cohen (1979), that formulating educational problems on a modest rather than grand scale has both psychological and practical advantages.

Problems that are modestly rather than grandly formulated are better matched to the limits of our information-processing capacities. Tighter analyses can be formulated when cause and effect linkages operate on a smaller scale, when critical evidence can be determined and detected, and when formal and informal experimentation can provide the feedback required for theory revision. When problems are formulated on a grand scale, it is very difficult to do the type of intensive analyses required to identify precisely how they arise and are sustained in the concrete practices of those involved in the situation.

Moving from analysis to intervention, Weick (1984) also argues that many social reform efforts have failed because they were too massive in scale. He compares, for example, the success of feminists' attempts to desex the language, with the relative failure of their attempt to desex legislation through the equal rights amendment to the United States constitution (p. 42). The reasons for failure are both task-related and psychological. Large tasks require more careful co-ordination than small ones and are thus more sensitive to breakdowns in timing, to political defections and to failure to achieve subordinate goals.

Psychologically, the chances of success are increased when massive social problems are recast into smaller ones, because problem-solvers are more likely to judge the challenge as within their capacities, and as less risky. As Weick puts it, aiming for small wins produces a more functional level of arousal for those involved.

> A small win reduces importance ("this is no big deal"), reduces demands ("that's all that needs to be done"), and raises perceived skill levels ("I can do at least that"). When reappraisals of problems take this form, arousal becomes less of a deterrent to solving them (Weick, 1984, p. 46).

Weick's advice, that massive social problems be recast into smaller problems which are better suited to solvers' cognitive and affective resources, is significantly different from that of writers who advocate a more holistic approach to social problem-solving. Cox (1981) attributes such an approach to critical theorists:

> As a matter of practice, critical theory, like problem-solving theory, takes as its starting point some aspect or particular sphere of human activity. But whereas the problem-solving approach leads to further analytical sub-division and limitation of the issue to be dealt with, the critical approach leads towards the construction of a larger picture of the whole of which the initially contemplated part is just one component, and seeks to understand the processes of change in which both parts and whole are involved (p. 129).

In advocating that theorists relate the part to the whole, Cox is assuming either that researchers have a theory of the whole to which the part can be usefully and substantively related, or that if they do not, a holistic, or what Lindblom (1979) calls a synoptic, approach to its construction is more useful than a more incremental approach. Weick and Lindblom would argue that both assumptions are problematic. Frequently, our large-scale theories are so vague and contested that they provide only the loosest of constraints on our local theorising. If we have a theory of the whole, then, of course, our component analysis should be coherent with it; Weick and Lindblom's point is that more often we will not know what it is a component of, and that a more incremental approach to problem-solving is a better path to the eventual discovery of large-scale theories than a more holistic approach.

My preference for tackling problems on a modest rather than a grand scale does not imply a preference for single-loop over more radical double-loop solutions to those problems. Where one sets the boundaries of a problem is independent of the degree of change required to resolve it. When double-loop changes are required, they may be more easily achieved when problems are modestly formulated, because it is easier in these contexts for both researchers and practitioners to learn precisely what is involved in achieving them, and to mobilise the resources required. In short, there are advantages for both the understanding and the resolution of social problems in formulating them on a modest rather than a grand scale.

INTERPERSONAL PROCESS AND PROBLEM-BASED METHODOLOGY

It was argued earlier that researchers who wish to contribute to the understanding and resolution of educational problems should reconstruct the theories of action of those involved in the problem situation, show how they might be implicated in the problem and, if necessary, develop an alternative theory of the problem which meets specified criteria of theoretical adequacy.

Only passing reference has been made to the researcher–practitioner relationship in which these methodological steps should be embedded. These interpersonal processes are critical to the success of PBM, because researchers alone do not resolve educational problems. They exercise their influence through the actions of those practitioners who control or can gain control over the relevant factors in the problem situation. A methodology which seeks to resolve educational problems, therefore, must describe the social relations of inquiry through which researchers can exercise this influence. What is needed is a relationship which fosters the commitment of all involved to a shared and adequate theory of the problem and of how to resolve it. If practitioners find the critique of the researcher compelling, and an alternative theory more attractive than their current one, then their intellectual commitment, rather than the unilateral control of regulators and researchers, should drive the change process.

People find arguments compelling when they have good reasons to believe they are true. Good reasons are those that have survived the test of people's doubts and criticisms. Practitioners may see researchers' reconstructions of their own theories of action as oversimplified and incomplete; researchers may believe that practitioners' objections to alternative theories are based on misunderstandings or defensive forms of reasoning. Until these objections have been resolved, those involved will not be committed to a shared theory of the problem.

The Nature and Conduct of Critical Dialogue

What type of relationship between researcher and practitioner facilitates intellectual commitment to the process and outcomes of problem-solving? It is easier to start by saying something about what is not involved in such a relationship than about what is involved. It is not primarily concerned with gaining the agreement of the practitioner to the researcher's views, because such goals prejudge the merit of those views and jeopardise the process of theory critique. What we are looking for instead, is a relationship between researcher and practitioner where the primary goal is warranted agreement about what counts as the best theory of the problem. By warranted agreement, I mean one that is justified by evidence and/or analytic argument, in contrast to one that is reached without attempting to test the warrant of each party's claims. The goal of warranted agreement suggests that the guiding value of the researcher–practitioner relationship should be inquiry and learning, through a process of mutual theoretical critique and bilateral control of the research process. I call such a relationship a process of critical dialogue (Robinson, 1989b).

Although many post-positivist educational and social science researchers call for a relationship of mutual critique and collaboration between researcher and practitioner, very little attention has been paid to the nature of such a relationship and to its practical possibility (Comstock, 1982; Lather, 1986; Lincoln & Guba, 1985). Those who have discussed the issue, frequently refer to the work of Jurgen Habermas on communicative competence and the ideal speech situation (Habermas, 1979, 1984), suggesting that he provides an outline of the type of dialogue required (Comstock, 1982; Marshall & Peters, 1985).

While there are considerable similarities between Habermas's communicative competence and my notion of critical dialogue, Habermas gives so little exposition of the precise nature of the relevant processes that I have turned instead to the work of Chris Argyris as a framework for my model (Argyris, 1982; Argyris, Putnam & McLain Smith, 1985; Argyris & Schön, 1974). Argyris has for the last twenty years been concerned with describing and creating interpersonal and organisational environments conducive to learning and problem-solving. The following model of critical dialogue (Figure 3.1) is based on what he calls Model Two, a generalised model of the interpersonal processes conducive to the achievement of these goals. Model Two is contrasted in Argyris's work with Model One, which forms the basis of the subsequent discussion of unilaterally controlling forms of dialogue. (I use the term critical dialogue rather than Model Two because it is more suggestive of its constituent properties.)

Critical dialogue comprises a value base, associated interpersonal skills, and the predicted interpersonal and task-related consequences of their employment. Under this model, the primary value which informs the interaction between researchers and practitioners is that of valid

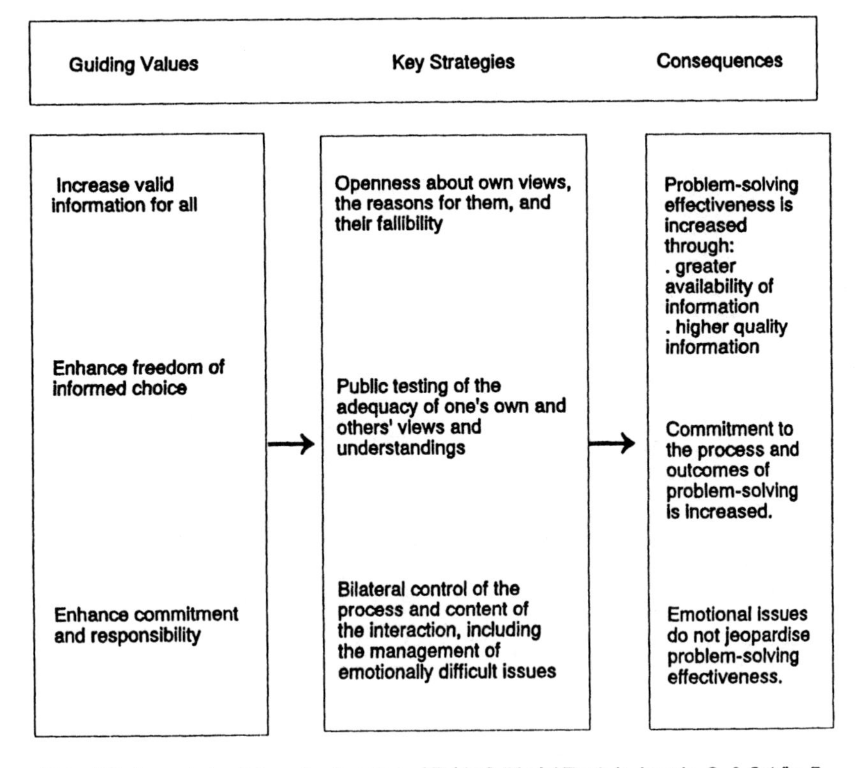

Note: This is a substantially revised version of Table 2 (Model Two), in Argyris, C. & Schön, D. (1974, p. 87)

FIGURE 3.1. A model of critical dialogue.

information. Decisions about what to say and how to say it are based on their implications for learning about the validity of one's own and others' views about the problem. Such views are treated as hypotheses to be tested, rather than as assumptions to be taken for granted or imposed on others.

The second value that informs critical dialogue is that of free and informed choice. Practitioners will not be intellectually committed to a course of action if they feel that they have been forced to participate, or that their participation is on the researcher's terms. The free involvement of those who have a stake in the process promotes the sharing of relevant information, more rigorous testing of views and more informed choices about what to do. A choice is informed when it is based on relevant information, where relevance is determined by each party's beliefs about the factors that would make a difference to the decision. The third value, that of internal commitment, is a consequence of adherence to the first two. Under this value, inquiry is conducted in ways that promote ownership of decisions and a sense of responsibility for their implementation, monitoring and possible revision.

The quality of dialogue between researchers and practitioners is judged by the translation of these values into behaviour, and not merely by participants' philosophical commitment, no matter how sincere. It is possible, therefore, for researchers who espouse collaborative, bilateral research practices to violate these values at the level of their theory-in-use. Adherence to the model outlined, therefore, requires the translation of these values into appropriate interpersonal skills.

There are three sets of interrelated skills involved in the conduct of a critical dialogue. The first set of skills concern openness; that is the ability to say what one thinks or wants in a way that increases the chance that others can do the same. Problem-solving is enhanced when all relevant information is put on the table, including that which is controversial, or potentially embarrassing to any of the parties. The second set of skills are concerned with testing; the point of openness is not just to hear a range of views, but to express them in ways that increase the chance that errors can be detected and corrected. When people say why they hold their views, as well as what they are, then others learn more about the positions being advocated and can judge the soundness of the arguments and evidence that are claimed to support them.

In the context of critical dialogue, testing incorporates a range of informal public checking processes, all of which involve exposing our perceptions, attributions and evaluations to the critical scrutiny of others. Attributions about others can be tested by disclosing them and saying how we arrived at them. ("I thought you were against this proposal because you didn't say anything at the meeting.") We can test the validity of our evaluations by disclosing the basis of our judgements and gaining others' reactions. ("I turned down your proposal because I believe parents will find it unacceptable. Do you see that as an issue?") We can test our assumptions that we understand each other by stating what we think they have said. ("Are you saying that you are not prepared to visit these parents' homes?") These testing and checking processes are informal versions of the more formal processes of theory adjudication that are central to any empirical endeavour.

The third set of skills concerns the way power and control are exercised by participants in the dialogue. The values of valid information, free choice and internal commitment are served by bilateral (or multilateral) rather than unilateral control of the content and process of the interaction. When there are inequalities of status and/or expertise, the achievement of bilateral control may require the more "senior" participants to facilitate the involvement of those who are more "junior", through explicit requests for their reactions, through assistance with the expression of their reactions, and through appropriate non-verbal skills. A highly articulate researcher who lacks the willingness or ability to help a practitioner articulate their view of the problem, and their reaction to that of the researcher, jeopardises both the process of theory adjudication and the development of commitment to a shared view of the problem.

The following imaginary conversation between a researcher (R) and teacher (T) illustrates how some of these skills help the parties reach a joint decision about whether or not the researcher will observe in the classroom.

Conversation	*Analysis*
R: I'd like to observe in your classroom to get some first-hand information about what the children are doing.	R discloses wish to observe and gives a reason.
T: I would have thought that my reports were detailed enough to tell you what was going on.	T challenges adequacy of R's reason.
R: They may well be, but I can't be sure unless I independently arrive at the same conclusions as you. If I don't, then we would need to discuss it further.	R acknowledges that T's reason may be inadequate, and elaborates it further.
Do you have some concerns about my coming into the classroom?	R facilitates expression of concern about classroom observation.
T: Well, I don't want anything done that could jeopardise my job here. I'm only in a relieving position.	T hints at a link between being observed and job security.
R: How do you think being observed might be harmful to you?	R facilitates expression of T's theory of being observed.
T: Well, if the principal knew that there were problems in my classroom . . .	
R: And you think he might find out through this research?	R checks own understanding of T's theory.
T: Yes, I suppose so.	
R: I certainly don't want to jeopardise your job, and I won't be talking to your principal without your permission. It's a dilemma isn't it—I'm saying I need the observations in order to help you solve the problem,	R lays out the apparent conflict between the two positions.
and you may be saying "How can I trust you to let you in to my classroom?"	R attributes concerns about trust to T and checks.

Continued

Conversation	*Analysis*
T: Yes, that's it. Maybe if I knew a bit more about what you want to do, and we could work closely together, I wouldn't feel so vulnerable.	T confirms R's understanding and suggests a possible solution.
R: Are you saying that if we discuss this in more detail you might feel differently about my presence in the classroom?	R tests own understanding of T's prior statement.
T: Yes.	
R: Well let's do that and then see how we both feel about going ahead.	R suggests that they test whether further discussion can resolve the issue to the satisfaction of both parties.

The skills of Model Two invite others to be part of making decisions and solving problems in the context of the best possible information. One consequence is that problem-solving is more effective because the parties learn about each other's views and why they hold them, including both their affective and cognitive reactions. Once these views are made public, their accuracy can be checked and their implications debated. For example, the researcher in the above example sought out the teacher's objections to classroom observation and once they were disclosed, their basis could be checked and procedures established to alleviate them.

A second consequence of critical dialogue is an enhanced sense of responsibility and commitment to the problem-solving process. This results from testing perceptions of what is at stake for each party, including both the costs and benefits of involvement, and of pursuing particular courses of action. For example, if after further discussion the teacher feels more confident that the researcher can be trusted, commitment to the research may correspondingly increase.

A third consequence of critical dialogue is that people make joint decisions about how to share potentially upsetting information without jeopardising problem-solving effectiveness. Understanding and resolution of long-standing problems frequently require discussion and disclosure of information that may be embarrassing to one or more parties. The value of seeking to maximise valid information requires that such information be included in the problem-solving process. At the same time, the principles of free and informed choice and bilateral control preclude one party from making unilateral decisions about whether to expose others to, or protect them from, any such embarrassment.

Unilateral Control as a Barrier to Critical Dialogue

Despite the high level of espousal of critical dialogue (Model Two), there is considerable social psychological evidence that actors do not behave accordingly, particularly in situations where they have something at stake and disagreement is anticipated (Argyris, 1982, p. 43; Bifano, 1989). In such situations, an espousal of bilateral control is contradicted, in practice, by highly skilled strategies of unilateral control, in which key assumptions are treated as self-evident and protected from critical examination. Protection may be provided by assertions about what is or ought to be the case, without provision of reasons or evidence, or by attempts to ignore or discourage opinions or information which question the views of the speaker (Argyris, Putnam & McLain Smith, 1985).

There is more involved in this incongruence than a simple failure to practice what one preaches, however, for Argyris's research also shows that these highly skilled unilateral strategies are the result of a theory of interpersonal interaction which is frequently tacit, and inconsistent with that which is espoused. Figure 3.2 describes the theory of unilateral control (Model One) which people frequently employ in tough situations, despite their espousal of the contrary.

The two guiding values of unilateral control are seeking to win rather than lose, and seeking to do so with the minimum of unpleasantness. The value of winning refers to getting what one wants, and maintaining one's view of the world, while protecting those wants and understandings from critical examination. It is not the desire to win that is problematic in this model, but the desire to do so in ways that protect goals from critical scrutiny and in ways that do not accord others an equal right to fulfilment of their (critically examined) goals and desires. The second value provides a constraint on the first, because winning must be achieved with the minimum unpleasantness to self and others.

The first set of strategies associated with unilateral control involves a failure to be completely open about one's views or the reasons for them. This failure may be attributable to deliberate manipulation or, more commonly, to a mistaken belief that they are self-evident and incontestable. Non-disclosure or incomplete disclosure leaves others guessing at the reasoning behind the actor's claims, and in a position of dependency rather than of equality.

A second set of strategies is concerned with private rather than public testing of one's views. Private judgements about the accuracy of one's beliefs and about how others are reacting result in unilateral adjustments designed to ensure that one's goals are achieved. Reading between the lines, strategising, second-guessing and mind reading are

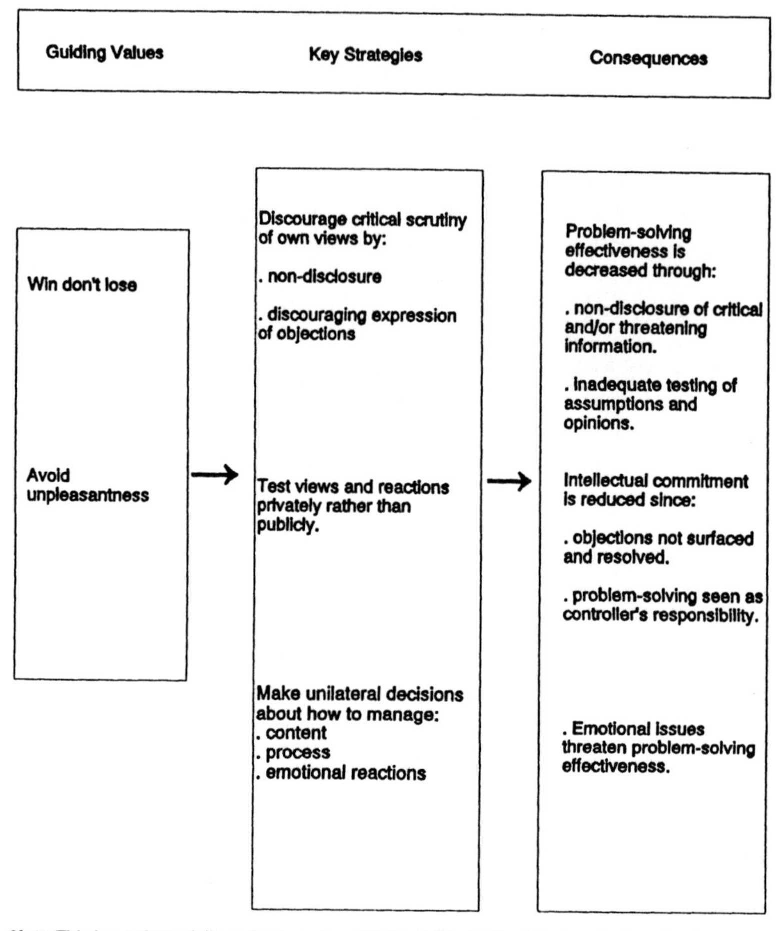

Note: This is a substantially revised version of Table 1 (Model One), in Argyris, C. & Schön, D. (1974, pp. 68-69)

FIGURE 3.2. Dialogue as unilateral control.

some of the everyday terms used to describe these private monitoring strategies.

The third set of unilaterally controlling strategies is difficult to separate from the first two. The value of winning is served by striving to keep control of both the process and content of key conversations. Unilateral judgements are made about how to interpret information, and about the goals to pursue and how to achieve them. In addition, unilateral decisions about the timing, sequence and content of messages are designed to keep one's own and others' (assumed) reactions within an acceptable emotional range. In short, unilateral control involves masterminding situations, whether from benevolent or malevolent intent, by taking control over goals, over how to involve others, over how to gather and interpret relevant information, and over how to manage people's feelings.

A desire to win and to do so in ways that minimise unpleasantness is non-problematic when our assumptions about the world are reasonably accurate and are shared by others. Unilateral control gets us into trouble, however, in the absence of either or both of these two conditions. When faced with tough problems which seem to inevitably arouse negative feelings, the injunction to reduce unpleasantness can lead to the sacrifice of problem-solving effectiveness through censoring critical information from the discussion. When those with differing views about the nature and resolution of the problem want a share in decision-making, unilateral control strategies can lead to an escalation of conflict, mistaken perceptions about people's interests, and pessimism about the ability of the parties to resolve the problem. Finally, when one party unilaterally manages the problem-solving process, others will feel a low level of commitment to the resulting decisions, either resisting them, or passively going along with what the controller wanted. In short, unilateral approaches threaten problem-solving effectiveness through both a reduction in quality information and through a reduction in the degree of commitment of those involved in the process.

More on Critical Dialogue

In the final section of Chapter 2, four criteria for evaluating theoretical adequacy were described and illustrated. The last of these criteria, improvability, described the qualities of a theory of action which foster or inhibit the detection and correction of error in the theory itself. These same qualities, adapted and expanded to suit an interpersonal context, form the basis of critical dialogue. In other words, the improvability criterion is used in PBM to both evaluate the theorising of practitioners, and to reflexively evaluate the discourse of researchers and practitioners.

It would be reasonable on first encountering the model of critical dialogue to object that it made excessive and impractical demands for disclosure and testing. After all, if researchers and practitioners attempt to share and test assumptions that they would normally have taken for granted, interaction will become laboured and self-conscious, rather than relaxed and spontaneous (Swanton, Robinson & Crosthwaite, 1989). The objection can be answered by making a distinction between adherence to the values of critical dialogue and employment of its skills. A concern for valid information and free and informed choice does not require constant disclosure, testing and inquiry, because there are some contexts in which full use of these strategies is unnecessary. Some contexts do not require highly precise communication because nothing much is at stake; other contexts do not require explicit checking because the roles, situations or past history of the parties have established shared understandings. It is true, of course, that sometimes one's assumption that an understanding is shared turns out to be false, and that there is no clear answer to the question "How does one know what can or cannot be assumed?" Suffice it to say, that an actor who is committed to the values of critical dialogue is far more likely to detect and correct mistaken assumptions than one who is not. Having discovered such mistakes, the actor would correct them by fuller use of the skills of critical dialogue.

The contexts in which explicit and more complete disclosure and testing are required, are those in which the views and interests of each party are unknown, or it is unclear how their differing interests can be reconciled, as in the dialogue between the researcher and the teacher who was concerned about being observed. These are, as we have seen, precisely the contexts in which unilateral strategies tend to overtake espousals of openness. More careful and deliberate use of the skills of critical dialogue is also appropriate when researchers and practitioners are operating in environments which bias interaction towards a model of unilateral control.

The Research Process and Critical Dialogue

In PBM the stages of inquiry are conducted as a process of critical dialogue between researcher and practitioner, directed towards mutual critique and education in the search for an adequate understanding and resolution of the focus problem. The implications of critical dialogue for the key tasks of PBM are spelled out in more detail in Table 3.1.

The table suggests that decisions about the accuracy of inferences about theories of action, the adequacy of rival theories and the success of alternative strategies are made in a bilateral or multilateral, rather than in a unilateral, manner. The description of the unilateral approach to the key tasks makes it clear that unilateral control can be exercised by either the researcher over the practitioner, or vice versa. From the point of view of fostering theory competition and shared responsibility, either form of unilateral control is equally undesirable.

Is There a Problem?

The description of key tasks presented in Table 3.1 assumes that both researchers and practitioners agree that there is a problem to be

TABLE 3.1. *Unilateral and Bilateral Approaches to the Key Tasks of Problem-based Methodology*

	Tasks	*Unilateral approach*	*Bilateral approach*
1.	Discover practitioners' understandings of the problem situation at both espoused theory and theory-in-use levels.	Accuracy of understanding is judged by the researcher or by practitioner.	Accuracy of understanding is negotiated in data-based manner.
2.	Critique practitioners' theories in terms of accuracy, effectiveness, coherence and improvability.	Persuade practitioners of the critique while protecting all parties from threat and embarrassment.	Encourage challenge of both the critique and the criteria on which it is based.
3.	Debate the merits of alternative understandings and theories of the problem.	Resolve disagreements through giving in or persuasion.	Resolve disagreements through public testing and evaluation of their implications. Make bilateral decisions about how to deal with unresolved issues.
4.	Learn how to employ agreed alternative in the problem situation.	Pedagogy is uncritical of the alternative.	Pedagogy is simultaneously concerned with learning and testing the alternative.
5.	Test the outcomes of the alternative.	Researchers or practitioners unilaterally decide whether the result is an improvement.	All relevant parties participate in the evaluation of whether the problem situation has improved.

investigated, even though they may initially hold quite different views as to its nature. Quite often, however, parties do not agree on whether or not a state of affairs *is* problematic, and if consensus on this issue is required before inquiry can get started, then PBM may be hostage to those whose interests are served by the status quo. Disputes over whether or not a situation is problematic reflect underlying theoretical differences about the implications of current practice and the predicted consequences of possible alternatives. Researchers who wish to improve current practice should seek to resolve disputes about the problematic status of such practices prior to investigating the alleged problem itself. Some might say that the resolution of such disagreement is a matter of politics or ideology, and hence not amenable to the intervention of researchers. Problem-based methodology suggests, however, that research can contribute to the resolution of such disagreements if they are conceptualised as theory competition and addressed in the context of a critical dialogue with those involved. Recent writing on unqualified school leavers (Trueba, Spindler & Spindler, 1989) can be used to show what might be involved. Some researchers (and practitioners) are convinced that high rates of unqualified school leavers reflects an educational problem; others believe that early school leaving might be an adaptive response to family circumstances and the sure prospect of further educational failure. Research on the nature and resolution of this "problem" will not impress the latter group unless it speaks directly to the reasoning that leads them to deny that it is such. If school drop-out is not an educational problem, then we would expect the self-esteem and life chances of such students to be no worse, if not better, than that of a comparable group of students who stayed at school. Empirical testing of this prediction would go some way to resolving the disagreement about the problematic status of early school leaving. Once such disagreement was resolved, then inquiry into the nature and resolution of the problem, as outlined in Table 3.1, could begin.

In summary, disputes about whether a state of affairs is problematic reflect theoretical differences whose implication can be unpacked and tested, in exactly the same way as can theoretical differences in the way a problem is formulated. PBM is not restricted to issues whose problematic status is already acknowledged by policy-makers and practitioners. Where the problematic status of a policy or practice is contested, however, it suggests that the goal of improvement of practice is best served by inquiry which attempts to resolve the disagreement, rather than by research, which adopts a disputed position as its starting point.

The Effectiveness–Acceptance Dilemma Revisited

In Chapter 1 it was argued that researchers who wish to improve practice are frequently caught in a dilemma between the acceptance and effectiveness of their theories of the problem. When their theories are critical of those employed by key players in the problem situation, practitioners may have difficulty accepting researchers' findings and recommendations. On the other hand, when researchers avoid or water down any such criticism, their advice may be more readily accepted, but less effective in resolving the problem.

Table 3.2 shows how this dilemma is irresolvable when researchers employ a unilateral theory-in-use. In a unilateral dialogue, a researcher seeks to both persuade practitioners of the correctness of his

TABLE 3.2. *Managing the Dilemma Between the Acceptance and Effectiveness of Research*

Unilateral version	*Critical dialogue version*
Researcher (R) believes practitioner's (P) theories of action are problematic.	R believes practitioner's theories of action are problematic.
R seeks to persuade P to own view.	R seeks to disclose and test this view of the problem with P. These processes will either confirm or alter R's views.
R believes that communication of own views, which are assumed to be correct, will be emotionally unpleasant.	R believes that the process of disclosing and testing may be unpleasant.
R wishes to avoid emotional unpleasantness.	R wishes to avoid making a unilateral decision about how to deal with P's emotions.
R therefore faces a choice between addressing the problem and emotional unpleasantness, or not addressing the problem and the emotional status quo. R must make a unilateral decision because disclosure of the dilemma risks emotional unpleasantness.	R can share responsibility with P for both addressing the problem and doing so in ways that improve rather than damage the relationship between them.

or her views and to do so with a minimum of unpleasantness. The more critical those views, the more the researcher is faced with a trade-off between these values (Johnston, 1990). Researchers may seek to resolve the dilemma through a variety of "soft-sell" strategies, which may avoid unpleasantness at the price of distorting the critique. Alternatively, they may seek to resolve it through hard-sell strategies which convey their views more forthrightly, but risk damaging the collaborative relationship. Either of these approaches is ineffective, and controlling, because the correctness of the researcher's views is assumed. The focus of the dialogue is the success of the persuasive process, not the validity of the views that are being communicated.

In a critical dialogue, the researcher frames the criticism as a hypothesis to be tested, rather than as a conclusion to be imposed. Such a framing immediately reduces the emotional threat of the communication as the possibility of reciprocal rather than unilateral influence is created. The other party can give their story and be heard, as well as challenge the views of the researcher. Researchers who are genuinely committed to public testing of their criticism may find that the emotionality that they had anticipated is more a function of their unilateral style of communication than of the content of their message. This is not to say that all emotionality is eliminated in critical dialogue.

Some criticisms are threat-inducing no matter how openly held. The point of critical dialogue is not to eliminate all such threat, but to eliminate the unilateral control that is responsible for some of it, and to jointly manage the rest in a way that does not jeopardise the accuracy and the completeness of the information available to participants in the problem-solving process. The result of critical dialogue should be that all parties feel that a tough issue has been tackled with honesty and sensitivity. The dilemma between problem-solving effectiveness and acceptance is reduced because the emotionality that results from unilateral control is eliminated, and the rest is jointly managed in a way that does not sacrifice problem-solving effectiveness.

Summary

In the first half of the chapter, PBM was defended against a number of possible objections. First, it was argued that a constraint inclusion account of problem-solving avoids the criticisms frequently levelled at narrow instrumental accounts of problem-solving, because it is not restricted to solution strategies which are compatible with current understandings of the problem.

A second possible criticism of PBM is that its focus on theories of action ignores, or at least underemphasises, the causal role of structural factors in educational problems. Two replies to this criticism were put forward. Given that any explanation of a problem can be deepened by asking how it itself can be further explained, it is important to contextualise such criticism by relating it to particular purposes and problems. The primary purpose of PBM is to improve practice, and the criteria of theoretical adequacy appropriate to this purpose have been spelled out in some detail. Thus PBM already incorporates procedures to test this criticism, through examination of the predicted or actual consequences of any such omission. The second reply to this criticism challenged the appropriateness of a clear-cut distinction between structural and psychological or social psychological explanations of a problem situation. If structural factors are implicated in a problem, they will be mediated through the understandings and actions of relevant agents. To focus on individuals and groups, therefore, does not necessarily imply that structural factors are ignored, even when agents are not conscious of their influence.

The final criticism of PBM concerned the modest scale of its problem formulation. In reply, it was argued that there are cognitive and affective advantages to modest rather than grand-scale problem formulation. Modest problems are better matched to our cognitive capacities, and are more likely to invite the engagement and commitment, rather than the pessimism and cynicism, of those whose efforts are essential to the success of the problem-solving process.

PBM incorporates a model of the social relations of inquiry that are conducive to the development of a shared and adequate theory of the problem, and a shared commitment to its resolution. Although the proposed model, called critical dialogue, is widely espoused, a body of social psychological research suggests it is not widely employed when either or both of the participants believe that a lot is at stake. In such situations, their interaction is more accurately described by a model of unilateral control in which each party seeks to impose his or her understandings, goals and preferred strategies on the other. Finally, the consequences of each of these modes of dialogue were discussed in terms of their implications for the conduct of PBM projects, and for the resolution of the acceptance–effectiveness dilemma discussed in Chapter 1.

References

Argyris, C. (1982). *Reasoning, learning and action.* San Francisco: Jossey-Bass.

Argyris, C. & Schön, D. (1974). *Theory in practice: Increasing professional effectiveness.* San Francisco: Jossey-Bass.

Argyris, C., Putnam, R. & McLain Smith, D. (1985). *Action science.* San Francisco: Jossey-Bass.

Bifano, S. L. (1989). Researching the professional practice of elementary principals. Combining qualitative methods and case study. *Journal of Educational Administration*, **27** (1), 58–70.

Comstock, D. E. (1982). A method for critical research. In E. Bredo & W. Feinberg (Eds.), *Knowledge and values in social and educational research*, pp. 370–390. Philadelphia: Temple University Press.

Cox, R. W. (1981). Social forces, states and world orders. *Millenium: Journal of International Studies*, **10** (2), 126–155.

Fay, B. (1975). *Social theory and political practice.* London: Allen and Unwin.

Habermas, J. (1979). *Communication and the evolution of society* (T. McCarthy, trans.). Boston: Beacon Press.

Habermas, J. (1984). *The theory of communicative action*, vol. 1 (T. McCarthy, trans.). Boston: Beacon Press. (Original work published 1981.)

Johnston, M. (1990). Experience and reflections on collaborative research. *International Journal of Qualitative Studies in Education*, **3** (2), 173–183.

Lather, P. (1986). Research as praxis. *Harvard Educational Review*, **56** (3), 257–277.

Lincoln, Y. S. & Guba, E. G. (1985). *Naturalistic inquiry.* Beverly Hills, CA: Sage.

Lindblom, C. E. (1979). Still muddling, not yet through. *Public Administration Review*, **39**, 517–526.

Lindblom, C. E. & Cohen, D. K. (1979). *Usable knowledge: Social science and social problem-solving.* New Haven: Yale University Press.

Marshall, J. & Peters, M. (1985). Evaluation and education: The ideal learning community. *Policy Sciences*, **18**, 263–288.

Open University (1986). Introducing education policy: principles and perspectives. (*Module 1. E333, Policy-making in education*). Milton Keynes: Open University.

Robinson, V. M. J. (1989). The nature and conduct of a critical dialogue. *New Zealand Journal of Educational Studies*, **24** (2), 175–187.

Swanton, C. H. M., Robinson, V. M. J. & Crosthwaite, J. (1989). Treating women as sex objects. *Journal of Social Philosophy*, **20** (3), 5–20.

Trueba, H. T., Spindler, G. & Spindler, L. (Eds.) (1989). *What do anthropologists have to say about dropouts?* Lewes: Falmer Press.

Weick, K. (1984). Small wins: Redefining the scale of social issues. *American Psychologist*, **39**, 40–50.

PART II

Using Problem-based Methodology

4

Analysing Educational Problems: The Case of a Professional Development Programme

The major thesis of PBM is that the resolution of educational problems is facilitated if they are understood in terms of the theories of action of those involved. While a thesis of this complexity is not proven by the evidence of only two case studies, the cases that are included in Part II provide a rich empirical base from which to examine the potential of PBM for understanding and resolving educational problems. Chapters 4 and 5 present the analyses of these cases, while the intervention process is described and evaluated in Chapter 7. Chapters 6 and 8 comprise methodological reflections upon the employment of PBM, including a systematic evaluation of the theoretical adequacy of the resulting problem analyses. These reflective chapters give a flavour for the process of doing PBM, and provide an opportunity to discuss some of the more technical and practical aspects of its conduct in the context of recent writing on qualitative research.

The researchers were motivated to do these cases by their long-standing interest in the interpersonal and organisational conditions that facilitated or inhibited learning and problem-solving. (The use of the plural "researchers" throughout the case descriptions acknowledges the fact that they were completed in conjunction with Michael Absolum, an educational psychologist specialising in staff and leadership development.) We were interested in such questions as: "What features of schools and of their leaders help them to identify, analyse and resolve important on-the-job problems?" "What strategies are used to resolve problems and how successful are they?" The principals were motivated to participate by the opportunity which the research gave them to examine and evaluate their problem-solving.

Our understanding of PBM was not nearly as sophisticated when we began in 1987 as it is now—hence, some of the comments in the retrospective reflections on the case studies. Nevertheless, we knew that our questions required us to locate practitioners who wished to collaborate with us in a long-term relationship which was focused upon their practice as inquirers and problem-solvers. The opportunity to locate such people came with two workshops on leadership style and problem-solving effectiveness which the author gave to an annual conference of the New Zealand Post Primary Principals' Association. The theme of the workshops was the relationship between leadership style and problem-solving effectiveness. Participants were invited to indicate at the end of each workshop whether or not they were interested in pursuing the topic further. Practical constraints of time and distance, and further information about the commitment required, eliminated many of the fifty or so principals who were initially interested. In the end, Carol and Tony, both new principals of large co-educational Auckland secondary schools, became our collaborators, largely through a process of self-selection. Carol, principal of Western College, wished us to focus on a concern she had about the school's professional development programme; Tony, principal of Northern Grammar, was concerned about his ability to sustain the democratic leadership style that he espoused. A more detailed account of our contracting around these focus problems is given in the appropriate section of each case.

PROFESSIONAL DEVELOPMENT AT WESTERN COLLEGE

Western College is a large West Auckland co-educational school with a staff of eighty-five, including four executive staff. It was founded in 1975 and is perceived by some of its community as emulating the English grammar school traditions of a much older central city boys' school. These traditions include an emphasis on academic and sporting excellence, and on high standards of uniform and behaviour. In recent years, Pacific Island immigrant families have settled in this semi-rural, predominantly European working-class area, with a resulting substantial increase in the ethnic diversity of the student population.

Carol took up her appointment as principal in September 1987, after being the school's deputy principal for six years. During her time as deputy, she had established a professional development programme which was considered a model for other schools. The programme was comprehensive, well organised and generously resourced. Carol saw the programme as pursuing three objectives:

1. The personal and professional development of staff so that they continue to learn.
2. A process of school development which ensures that it remains responsive to change.

3. The development of a consultative school, in which teachers take as full a part as possible in decision-making, and so contribute to their own and the school's development.

The programme had two main thrusts. The first was the promotion of staff involvement in learning about curriculum or management issues through discussion with colleagues or visitors during departmental, full staff or special interest meetings or courses. Professional development opportunities were also provided by encouraging staff to observe or participate in selected programmes in other schools. The second thrust was a formal process for the supervision of staff with management responsibilities. This was called the Professional Development Consultation Cycle (PDC) and consisted of two or three interviews a year between one of the four executive staff and a middle manager, such as a Dean or Head of Department (Stewart & Prebble, 1985). The purpose of the consultations was both to support staff in their management roles and to hold them accountable for the functioning of their departments. Carol hoped that as middle managers became more confident in their role, they in turn would set up PDC consultations with the members of their own departments.

The Research Contract

We developed a contract with the principal to investigate her concerns that the professional development programme was not having the desired impact on teaching and management practices, and that a few staff continued to be resistant to its operation. It is important to note, however, that at this stage Carol did not perceive these problems to be particularly serious or urgent. She herself was engaged, through her postgraduate studies in educational administration, in research on teacher professional development, and she thought it would be interesting to gain the additional perspective of outside, university-based researchers. Two and a half years after we first entered into this contract, Carol presented us with a paper entitled "Collaborative Research: A Subject's Perspective". This is how she described her reasons for suggesting that we focus on the professional development programme:

> When approached by the researchers to identify an institutional problem which was important to me and could be an acceptable focus for research, I chose to reveal the vague concern I felt in relation to staff response to the professional development programme. I was uneasy about the impact the programme was having on classroom practice and the issue was extremely important to me because school resources, personal study and my

> professional credibility had been invested in the programme. To be honest, I did not believe that investigation would reveal a major problem but hoped that the research might confirm this view and also present an opportunity to learn some practical and relevant skills (Cardno, 1990).

When the research started, the PDC programme had been operating for about four years, and several previous attempts to increase its acceptance had been unsuccessful. We suggested to Carol, therefore, that the following research questions form the basis of the investigation:

1. Is there a problem? Are Carol's concerns about the impact and the acceptance of the programme substantiated by other forms of evidence?
2. If Carol's concerns are confirmed, how can they be explained?
3. Why can't Carol solve this problem? What dilemmas does she face in attempting to do so?

The Research Approach

The first research question reflects our wish to test rather than uncritically accept the principal's concerns about the programme. It signals our belief that collaboration implies debate about independently derived theoretical positions, including about the existence of problems, rather than an alliance with any of the parties involved in a problem situation. We answered this first question by seeking the independent opinions of different stakeholders rather than by conducting a formal evaluation of the programme's effectiveness. We decided against this latter course because, if a problem was confirmed by these perception checks, we wanted to devote the majority of our scarce research resources to its explanation and resolution, rather than to its identification.

Assuming that the nominated problems of low impact and resistance are confirmed, PBM suggests that the second research question should be answered by testing the links between these problems and the theory of action of the professional development programme. A theory of action of a programme is made up of the same components as that of an individual; the constraint structure, the practices which satisfy those constraints, and their consequences. In solving the problem (in the neutral sense) of how to design and implement a professional development programme, the principal and staff of Western College would have set or adopted a number of constraints on solution adequacy. These constraints may have been conceptual (e.g. activities which do not count as professional development are excluded from the programme), value-based (e.g. activities which violate values of collegiality and support are ruled out), or practical (shortages of time and money limit what can be offered). Explaining the problems of low impact and staff resistance, therefore, involved identifying the constraints and strategies that had led to these outcomes. Answering the third research question involved an analysis of the leadership and problem-solving style of the principal herself, in an attempt to explain why, despite her commitment, Carol had found these problems intractable.

The data required for these tasks were collected between September 1987 and March 1988 when the feedback and intervention phase of the project began. During this time five interviews were held with the principal to establish the research contract and to learn about her views on professional development. Her actual practice was studied through recordings of two PDC consultations and through her leadership of two executive, three senior staff and one full staff meeting. Tape recordings and field notes were also made of the proceedings of a school development day attended by most staff, of a full-day course for Deans on counselling and communication, and of a full staff meeting devoted to professional development activities. Perceptions of the programme and of Carol's leadership style were obtained through individual interviews with the executive staff (deputy and assistant principals) and with six other staff chosen to represent a range of views about the professional development programme.

Is There a Problem?

We quickly discovered that the programme was not operating as intended. Time and other pressures had meant that executive staff had had difficulty keeping up with their consultation schedules and few middle managers were using the PDC procedure with their own staff.

> *Exec. member*: It's not been a high priority. It's not something that's essential to the day-to-day routine, so it tends to get left to the occasional prompter. All of a sudden we're asked if we've done any [PDC], and we all rush around and chase our people up to have another go [WCTR9, 2.11.87].

Our interviews with staff revealed mixed evidence about the impact of the two aspects of the school's programme. Some staff could describe ways in which visits to other schools, involvement in curriculum associations, and particular in-service courses had influenced their teaching practice, but very few could describe ways in which the second more supervisory component of the school's professional development programme (the PDC) had led to any improvements in their own practice. This was partly a function of the type of conversation that occurred during these consultations. Staff's descriptions of their content suggested that their focus was more often on the exchange of information about forthcoming events, staff changes and resource priorities than on an examination of the staff member's teaching and administrative practice. The degree of impact of the PDC was also attributed, by some staff, to the age and experience of the recipient. One executive member believed that the PDC process was most effective with staff newly appointed to management positions, and that it made little impact on staff who were more reluctant to make changes. This view was partly confirmed by an older Head of Department:

> *HOD*: ... I'm sure it would have [an impact] on some younger people who haven't seen enough things perhaps. But I don't think it's had a great effect on me. I'm sure it probably is a weakness of mine, rather than of the system [WCTR8, 2.11.87].

Although these concerns focus on the PDC, the usefulness of the less formal professional development activities was also questioned by a few staff. They were cynical about whether management were serious in their claim that one of the purposes of these activities was to enable the staff to contribute to decisions about school development. These comments from various staff were logged by the researchers during a workshop on "Meeting Needs" that was held during a school development day:

> *Staff A*: ... We have to feel good about ourselves so that we can meet kids' needs. That's one of the functions of this day—it's a safety valve.
>
> *Staff B*: ... We are here to rubber stamp the changes that Carol already wants. I don't think the school operates smoothly and efficiently [as claimed by Carol in her prior address to staff].
>
> *Staff C*: ... I think we need to look at the lists [of suggestions] we have come up with and then take them to the whole staff.
>
> *Staff D*: ... But the decisions are already made when it goes to the whole staff. You get a sense of powerlessness—I'm not here to be a rubber stamp [FN, 2.10.87].

Having validated Carol's concern that the professional development programme was not working as she would have liked, we then set out to uncover the major beliefs and assumptions that guided the programme. These are presented in the next three sections as we consider how professional development opportunities at Western College were selected, designed and evaluated. In the subsequent section, we show how these beliefs and assumptions led to the problems of impact and acceptance.

Key Beliefs and Practices in Professional Development

The Selection of Activities

The professional development programme was organised by the Professional Education Group (PEG), a committee of staff which included the principal, and which was chaired by a senior staff member. The committee attempted to select activities that met the professional development "needs" of the school and its teachers. Some professional development needs arose because staff required support in the introduction of new curricula or administrative practices. For example, the decision to introduce longer home-room periods was accompanied by training workshops designed to help form teachers adopt a more pastoral role with their form groups.

Other professional development needs were assessed by the popularity of staff requests for particular types of activity. Staff were regularly asked to nominate possible activities, and more formal surveys were also taken of staff preferences. One of the jobs of committee members was to bring these suggestions to meetings where future activities were being planned. In the following extract, the Professional Education Group deliberates the selection of topics for the following year.

> *Chair*: (. . .) I was having a bit of a think about this, this afternoon, and there are certain areas which I think we could probably tackle next year. The question of induction of new staff. (. . .) There is also your suggestion Carol, which we have been doing over the last few years' of having some administrative courses, particularly for Deans. (. . .)
>
> *Carol*: Some of the other things that we could look at came out of the professional development day and the forum discussions [open meetings of the whole staff]. At the end of the sheet [summarising staff preferences] is five or six suggestions of areas which we could look at perhaps short term. (. . .) And of those there, probably the most important would be something along the lines of teaching mixed ability classes [WCTR23, 8.12.87].

These selection practices tell us that professional development activities are selected because they support innovations or because they are desired by Carol or the staff. We found little evidence that these suggestions were critically evaluated in terms of how they might improve current school or teacher practices. A staff request for a workshop on stress management, for example, reflects implicit assumptions about the nature of the problem for which a workshop on stress management is thought to be the solution. Different assumptions about the nature of the problem would yield different beliefs about possible solutions and hence differing requests. The committee showed little awareness of such assumptions, possibly because once they interpreted staff requests as "needs" they saw their role as trying their best, within resource constraints, to meet them.

The Design of Activities

The wide variety of professional development activities offered at Western College were mostly voluntary, though the limits of staff choice over attendance at some activities such as "teacher-only" days was always somewhat ambiguous. Given the (mostly) voluntary nature of staff involvement, Carol and the PEG were concerned to make the programme as attractive as possible to staff. This concern affected both the selection and the design of the various activities.

Carol and the committee believed that the staff were, on the whole, bored by lengthy presentations and reluctant to speak their mind in a large group of staff. They attempted, therefore, to maximise the amount of time staff spent in small groups, and to organise such time as efficiently as possible. This frequently involved the selection of a task for small group discussion, the preparation of relevant worksheets, and the appointment of staff within each small group to act as recorder and chairperson. Attention was also given to the way small-group discussions could be reported back to the whole staff group, so that staff could learn how their contributions were being used.

The following example illustrates some of the committee's assumptions about how to design effective and efficient professional development activities. During a staff workshop on school development held as part of the 1987 School Development Day, staff were asked to list the things the school "does well" and "the problems we have". The lists were returned to the Professional Education Group who, conscious of staff's desire for follow-up, wanted to report back to staff as soon as possible. The committee reduced the various lists to a manageable size and then asked staff for further reactions at the next general staff meeting. They prepared the following handout (reproduced opposite) as the basis for staff discussion on the problems that staff perceived in the school.

At the staff meeting, staff worked in small groups to rate each item as a major or minor problem on a 1 to 5 scale, and these ratings were then

Staff Open Forum—School Development Review

Below are a list of matters which need immediate reviewing. To what extent do you see these as problems? You will have your chance to air your views at the open forum meeting tonight after school. These matters have been raised by individual staff during the group discussion on school and staff development.

1. Defining a dress code for staff.
2. The availability of the hall—maximising its use
3. Making a staff member responsible for notifying staff about any functions held in the staff room.
4. Improving communication with ancillary staff.
5. An explanation of how PRs and promotion are obtained within the school.
6. Litter problem.
7. Reducing paper war.
8. Support of staff by senior staff when conflicts arise with students.

resubmitted to the Professional Education Group. Subsequent negative comments about this activity at a senior staff meeting suggested that it was not participation *per se* that staff wanted, but participation that made a difference to school decision-making and that led to greater understanding of the issues involved.

> *Exec. member*: I think we've done a fair amount of sitting around and talking this year, and I think that probably the staff are going to get a little bit cynical if we do it too often.
>
> *HOD*: I don't mind sitting around discussing, as [executive member] said. Provided it's not like it was on Monday night, with a whole list of things, and you sit and discuss them, but you don't know what. There is no tie in at the end. For example, the staff dress standards, litter—it was throw everything in, but it was very frustrating if you don't know what the flipping heck is going to happen after you've sat around and talked all that time about it [WCTR15, 5.11.87].

In their efforts to promote staff participation under the constraints of time and large numbers, the committee sometimes used processes that left staff bewildered about the meaning and significance of the task they were involved with. Staff moved efficiently through a sequence of activities that yielded a prioritised list of staff concerns, but there was little indication from those same staff that they understood the meaning or significance of either the original staff concerns or of the summary presented on the handout.

The design of the more formal PDC process was partly left to the two individuals involved in each consultation, and was partly prescribed by executive guidelines about what should be covered in each term's meeting. In our initial interviews with staff, several Heads of Department complained about how they had been asked to prepare a list of their staff's qualifications and experience prior to their most recent consultation. They could not understand how such an exercise contributed to their professional development, and so saw it as unnecessary paperwork. One Head of Department, however, sensed that it had a more important but unacknowledged purpose.

> *HOD*: (. . .) well, take the last [PDC] for instance, when you're asked to list your staff with their qualifications. That seems a petty administrative thing, which is time wasting, whereas if it's meant to mean, "Have you checked on your staff and asked them about this, that, and the other thing", then it has some meaning. A lot of it, I think, the phraseology or the way it's set up, it's intended perhaps to do something, but it does something different.
>
> *VR*: Right, so you're saying if it is meant to look at the way you are acting as HOD, then it needs to be somewhat more direct about that rather than asking you to list your staff?
>
> *HOD*: Oh yeah, that's right.
>
> *VR*: Have I got that right?
>
> *HOD*: There are things like that where, I suppose everybody needs checking up and they need check points and ways of doing it. But that should be the stated reason, rather than the peripheral [one] [WCTR8, 2.11.87].

The comments we received about this activity suggested that the executive were having difficulty being "up-front" about their desire to check the way heads of department managed their own staff. This indirectness, together with the executive's reluctance to make PDCs compulsory for middle managers and their staff, suggested a reluctance to openly pursue the accountability function of the professional development consultations.

The Evaluation of Activities

Individual professional development activities at Western College were most commonly evaluated by questionnaires to staff asking about their satisfaction with any particular activity and about ideas for improvement. The programme as a whole was evaluated by the committee against criteria of comprehensiveness and staff involvement. In his report to the Professional Education Group on the year's activities, the chairperson listed the variety of activities that had been provided at staff meetings, workshops and at departmental level. The group expressed pleasure at the comprehensiveness of the programme, but did not raise the issue of its impact on the teaching and administrative skills of the staff [FN, 8.12.87].

Explaining the Problems of Low Impact and Resistance

To what extent are the patterns described under selection, design and evaluation implicated in the problems of impact and staff resistance which Carol outlined at the beginning of our project? The following explanations were formulated by the researchers and repeatedly checked and revised in discussion with Carol and her executive staff.

The patterns we identified suggest that the professional development programme was designed to promote staff participation in activities which met their perceived needs, which were efficiently organised, and which they found satisfying. The strategies used to promote staff participation and satisfaction, however, simultaneously restricted the impact of the programme upon teaching and learning processes within the school. First, while staff were consulted widely about the selection of activities, and about their evaluation, these consultative processes produced a plethora of suggestions and reactions, rather than a critical appraisal of staff and school practices. The connections, therefore, between the variety of development opportunities that were offered and the practices that they might influence were often vague and left to chance. Second, Carol and the members of the committee believed strongly that the programme needed to be collegial, supportive and non-threatening to gain the enthusiastic participation of the staff. She described the programme and staff's reactions to it this way:

> *Carol*: ... I haven't really identified some of the metaphors and things we use, but certainly it's I think you'll find people are culturally very comfortable [with it]. It's not seen as a threat to their competence, they're not given PD because they can't do something; it's not remedial or a skill deficiency that we're pointing out. It's very much a stimulation, enthusiasm building kind of attitude to PD [WCTR30, 7.3.88].

Threat is felt when issues central to one's self-esteem are addressed, such as one's competence in important teaching or management roles. People can be protected from threat by avoiding such issues or by addressing them indirectly so that individuals can choose whether or not to connect the discussion to their own practice (Argyris, 1985, pp. 72–77). This was the main way in which the staff at Western College had attempted to run a programme which was perceived as non-threatening and supportive. The professional development activities were not selected on the basis of public critical appraisal of current teacher and school practices, they were not designed to impact directly on these practices, and they were not evaluated in terms of degree of learning or improvement. Similarly, none of the activities of the professional development programme were obligatory and the PDC itself could, in practice, be avoided by those who put their mind to it.

One can object to this analysis by arguing that a professional development programme can have an impact even though it does not directly address and evaluate current school and teacher practices. After all, teachers are intelligent professionals and their mere exposure to different ideas and activities may lead them to think about and improve their own practice. Why address this goal directly and possibly raise feelings of threat and insecurity, when a more low-key approach aimed at gaining staff participation may in the long run lead to self-education and self-improvement?

The answer to this objection involves consideration of the organisational consequences of threat avoidance. A professional development programme which seeks to be supportive and non-threatening will have difficulty influencing staff who rightly or wrongly are believed to be threatened by criticism. Frequently, it is those staff who are privately believed by their colleagues to be most in need of professional development who show no signs of voluntarily making the changes that others believe are desirable.

Similarly, the desire to avoid threat can lead to superficial analyses and to superficial solutions of school problems. During a professional development activity for Deans, staff addressed the problem of the number of students being sent out of class by teachers. Two years earlier, Heads of Departments had found the referral of such students by their own teachers too disruptive of their own work, so a procedure was set up whereby such students reported to a duty Dean. Now it was the Deans who were complaining that they were overloaded with referrals. A new set of procedures was designed which required Heads of Departments to take more responsibility for such referrals and to provide support to their own teachers. At no stage in this sequence of problem-solving was there an analysis of *why* some teachers were making so many referrals. This preventive focus would have involved discussion of teachers' instructional and management skills, and of the support they were currently receiving. Such a discussion may have been threatening for the teachers and Heads of Department involved, but the cost of avoiding it is probably another cycle of short-term solutions to the problem of teachers sending students out of class [WCTR 28, 1.3.88].

The wish to avoid threat presented the executive of Western College with a dilemma, because their roles also required them to address accountability issues such as teacher and administrator competence, which are frequently threat-inducing (Figure 4.1). They attempted to resolve this dilemma by either dealing with threatening issues outside the professional development programme, dealing with them indirectly

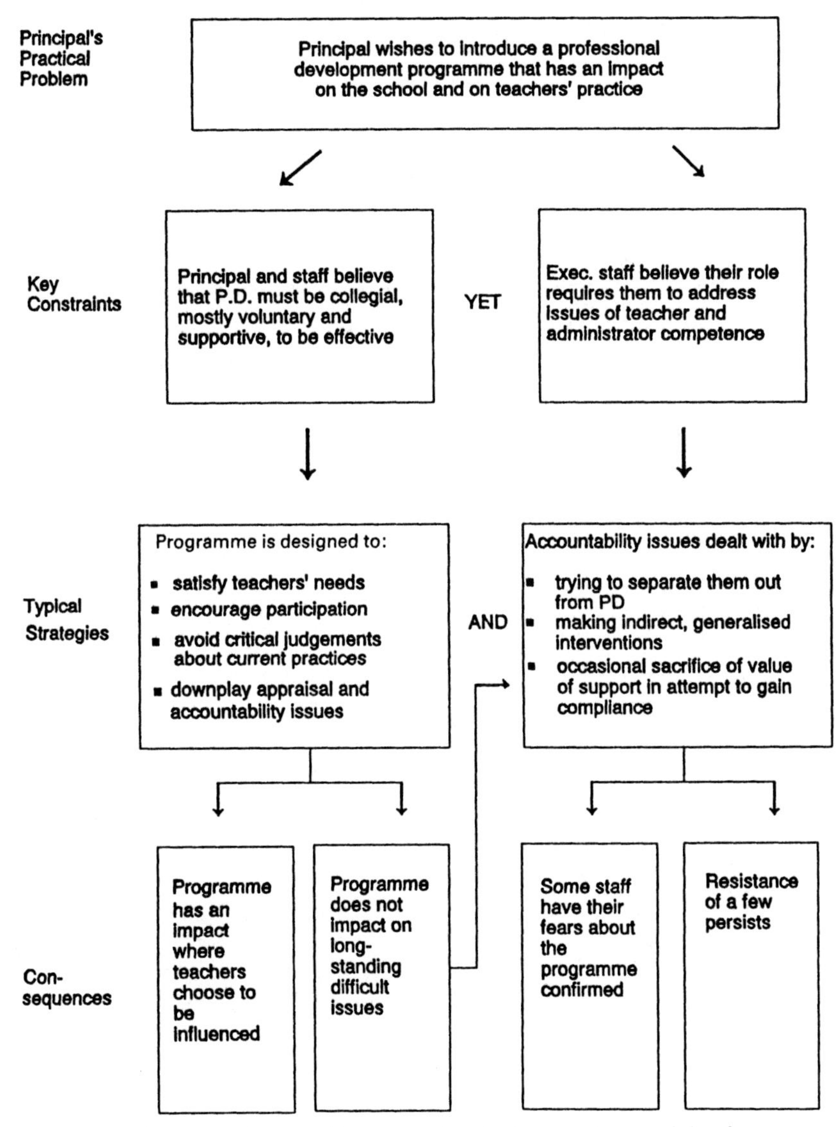

FIGURE 4.1. Analysis of problems associated with a professional development programme.

inside the programme, or occasionally forsaking the values of support and threat avoidance in the interest of teacher compliance. As is shown below, all of these responses compounded rather than resolved their dilemma.

When management dealt with tough issues outside of the programme, they contributed to the perception that professional development was a "soft option", incapable of addressing issues which were central to the functioning of the school. Part of the resistance of a few staff to aspects of the programme was attributable to this perception that it was an optional extra, secondary to the real job of teaching and running the school [WCTR5, 21.10.87].

When management did not succeed in separating out these tough issues, their portrayal of the professional development programme as supportive and non-threatening lost credibility. Some staff smelt an "accountability rat" lurking in the programme and thereby felt justified in their prior suspicions about the purposes of professional development. This suspicion was expressed by Heads of Departments in their complaints about the request to list the experience and qualifications of staff, and the executive staff acknowledged that they had deliberately downplayed the accountability functions of the PDC.

Management's indirect strategy was evident in the way members of the professional education group, including the principal, discussed the problem of how to encourage staff to use more effective and positive disciplinary strategies. Resources relevant to the perceived problems were to be "republished", "drip-fed" or "waved around" in an effort to persuade classroom teachers to re-examine their own practices [WCTR23, 8.12.87]. Where such material was consistent with staff's current beliefs, and they had a desire to do so, staff attempted to adopt some of the suggestions made. Where they did not see it as relevant, or where the assumptions implicit in the material (for example, that teachers should take responsibility for preventing certain sorts of pupil disruption) conflicted with their own beliefs, then such influence strategies had little impact.

When the executive felt compelled to solve an organisational problem, and such "softly, softly" approaches failed, their dilemma intensified. In cases where compliance was no longer considered optional, the value of staff support was sacrificed and more potentially threatening strategies adopted. In the following example, drawn from a tape-recorded PDC, the principal attempts to persuade a recalcitrant Head of Department to comply with several procedures which she believes are essential to effective staff management. She knows that he has not complied in the past and that he has serious reservations about the PDC, but she ignores this information in her attempt to gain his compliance.

Carol: Right, now our List A [first year] staff. That was something I felt urgently we had to discuss.

HOD: Yes.

Carol: What I really wanted from you was a little report on each one of them.

HOD: Mmmm.

Carol: And I would like that if possible by the end of the month.

HOD: Alright, yes.

Carol: Just a brief comment.

HOD: Mmmm.

Carol: On Peter, one on Michael, and who else have we had who has come in? Don't worry about Jill and Jo, no. That's fine.

HOD: The others—

Carol: Who have I missed now . . . they're List B, so that's alright. It's just those two. So if you could just do me a little report on each of them.

HOD: Certainly.

Carol: And the other urgent thing is your own department's PDCs. That's one of the things we were trying to encourage. . . .

HOD: Mmmm.

Carol: . . . HOD's to get involved in. Now if you've got Paul who is going to assume some of your responsibilities and perhaps actively help with the resources and things like that, that should give you a little bit of time to meet—

HOD: Mmmm. Certainly.

Carol: Even if it's just once in the year . . . [WCTR19, 23.11.87].

The HOD is told in quick succession to provide reports on his List A teachers, to take urgent action to implement a PDC process with his own staff, and that this latter request is reasonable since he now has relief from a newly appointed senior teacher. These strategies are controlling because the principal asserts the importance of compliance without saying why she believes such compliance will assist staff management, she does not solicit the HOD's views despite her awareness of his doubts about the PDC, and she fails to explore his reasons for not complying so far. She may gain compliance, but the principal and

HOD may be no closer to a shared understanding of the nature and value of the PDC process. Later comments by the HOD did show that he still held the same reservations about PDC, but saw this particular conversation as one in which he had no option but to agree [WCTR62, 20.11.89]. For staff like this Head of Department, who have reservations about PDC, this type of discussion may alter public expressions of agreement, but will leave their private resistance to the programme unaddressed.

In summary, the professional development programme at Western College did not have the expected impact on management and teaching, because it was not designed to influence those processes directly. Instead, it sought to encourage staff participation in a wide range of professional development activities, which they could use as opportunities to reflect upon and alter their own and the school's practices. A more direct connection between the professional development activities and teacher and school practices was precluded by a desire to conduct a supportive and non-threatening programme. This meant that critical discussion of teacher and school practices was difficult, and that the executive were unable to address accountability issues while remaining consistent with the ethos of the programme.

Explaining the Persistence of the Problems

Given the commitment to professional development at Western College, it is important to ask our third question about why the principal and her executive had been unable to resolve the problems that they had perceived. Three answers to this question are suggested, ranging from current policy on professional development, to the academic theories of professional development which influenced the principal, to the leadership style which she demonstrated in addressing problems that she perceived in the programme. All three influences served to engender and reinforce the belief that effective professional development was supportive and collegial, and that being supportive in turn involved the avoidance of both threat and issues of accountability.

The Policy Context of Professional Development

In 1980, when Western College first embarked on the task of providing for staff development, the concept was not associated with the ideas of accountability and school self-management that were to become so important under the subsequent 1989 reorganisation of the administration of New Zealand education (Department of Education, 1988). At the time, the concept was largely associated with the provision of centrally funded regional or school-based in-service courses for teachers (Cardno, 1988). It is not surprising, then, that teachers saw professional development as something that happened to them on special courses, and that they generally expected to be directed to participate and to take little responsibility themselves.

Given these expectations, Carol set herself some very modest initial goals. If she could get teachers who did not expect to be involved in, or responsible for, their own professional development to experience and to feel positive about its possibilities, she would be satisfied. In the educational climate of the early 1980s this goal was relatively non-problematic, because there was far less emphasis at that time on being accountable, both in terms of resources and teaching quality, than there was after the 1989 reorganisation of the administration of New Zealand education. Gaining teacher participation, therefore, was a first step, and she turned to the literature on professional development for some ideas about how to achieve this.

The Professional Development Literature

Much of the literature that Carol turned to stressed that the area of professional development for teachers was fraught with political, moral and practical difficulties. It warned her that teachers would be resistant to any invasion of their autonomy in the classroom, and that if they were going to be involved in any programme involving collegial oversight and supervision, "they must be assured that the relationship will be a helping and sustaining one, rather than a dangerous exercise in exposing one's weaknesses" (Prebble & Stewart, 1983, p. 52).

One way in which supportive relationships could be developed, according to this literature, was by clearly distinguishing between the formative and the summative purposes of any evaluation. Formative evaluations are associated with developmental and supportive encounters between two colleagues designed to assist one party's self-development, while summative evaluations provide an information base for a variety of threat-inducing personnel decisions such as confirmation of appointment, promotion and termination (Codd, 1983, p. 67). Other writers seek to reduce further the threat involved in supervision and evaluation by eliminating the judgemental element involved. Crane, for example, advises supervisors to provide descriptive feedback to their colleagues and to avoid making judgements about the merits of what they have observed (Crane, 1975).

In short, the literature that Carol read stressed the importance of treading carefully, the importance of developing a programme that was not accountability driven, and the need to emphasise the developmental rather than the judgemental purposes of the programme. This advice was one-sided because it did not acknowledge that judgements are an inevitable part of teacher–teacher and teacher–administrator interaction. If professional development programmes are supposed to be non-judgemental, then these judgements will simply be driven underground. In addition, the advice did not acknowledge the way senior staff were increasingly being held accountable for the quality of teaching and learning in their schools. When staff disagree about issues of quality, threat is likely to be felt and a variety of defensive reactions displayed. If threat avoidance is a central concern, learning and problem-solving are likely to be sacrificed.

Carol was unaware of the limitations of this advice until we started to challenge some of her assumptions about the conditions required for effective professional development. She had set out to establish a programme consistent with her reading, and when she later experienced its inevitable limitations, she drew on her own leadership and problem-solving style for coping strategies.

The Principal's Problem-solving Style

The third and perhaps most important explanation for the persistence of these problems is linked to Carol's own problem-solving and leadership style. We explained at the beginning of the case how Carol did not see the problems of impact and resistance as serious. This explanation, however, raises further questions about why she misperceived the problems in this way. After analysing and discussing numerous relevant transcripts of meetings, we came to believe that unilateral features of Carol's style prevented the recognition and resolution of these problems.

Carol was enthusiastic about professional development and saw it as part of her role to engender enthusiasm in others. One way she did this was to make frequent positive statements to staff about the professional development programme. In the final meeting of the Professional Education Group for 1987, for example, she evaluated the programme as follows:

> *Carol*: They [the objectives for professional development] are all ideals of course, but I think we've got fairly close towards achieving them. (...) And in conclusion, the role played this year by the Professional Education Group, (...) has exceeded the expectations I held for the establishment of a valuable and continuing professional development group for the school. I know for a fact that this school has for the last four years been in the vanguard of schools concerned with promoting the personal and professional development of their staff... [WCTR23, 8.12.87].

The principal conveyed a certainty about the worth of the programme and did not ask other members of the group whether they agreed with her evaluation. Her intention was not to gather information from them, but to convey her genuine pleasure at what she believed was a job well done. In terms of the criteria that she used to judge the programme, namely variety of opportunities offered, participation by staff, and efficient and responsive provision, the 1987 programme had been a great success. Given Carol's widely acknowledged enthusiasm for and commitment to her programme, her position of power, and her failure to facilitate others' views, these unilateral positive evaluations constrained other staff from expressing more critical opinions about the programme.

When Carol did perceive or learn of problems in the programme, the problem-solving process was curtailed by her tendency to "have the answer" before she or others had had a chance to think things through. Several colleagues commented in their interviews on her decisiveness and the way it resulted at times in oversimplified responses to complex issues. This response was explained by Carol's belief that it was the principal's role to be decisive and to act, and one way of meeting this imperative was to have answers when others were uncertain. As staff perceived Carol responding this way, their expectations in turn trapped Carol in this role. Linked to this tendency to "have the answers" was a more general tendency to make strong assertions about what was or should be the case, in ways that others found difficult to challenge. We previously illustrated this controlling style with an extract from Carol's PDC interview with a resistant Head of Department. Further evidence was obtained from the transcript of a meeting of the executive staff which we had analysed in response to Carol's request for feedback on her style as chairperson. Our analysis showed that while Carol negotiated procedural matters and was open to challenge, she more frequently made strong uncompromising statements and unilateral declarations about the nature of problems and solutions [WCTR37, 19.7.88].

Carol's decisiveness and unilateral evaluations were tempered by her own acceptance of the professional development value of threat avoidance. When she was faced with a difficult issue of accountability, this value placed her in the same sort of dilemma that I have argued reduced the impact of the whole professional development programme. Figure 4.2 shows how she attempted to resolve this dilemma in the context of her concerns about the Head of Department who featured in the previous interview extract [WCTR19, p. 28]. She oscillated between indirect strategies that were unilaterally protective of herself and her Head of Department and more forthright controlling strategies such as those seen in the interview extract. The result was that in this

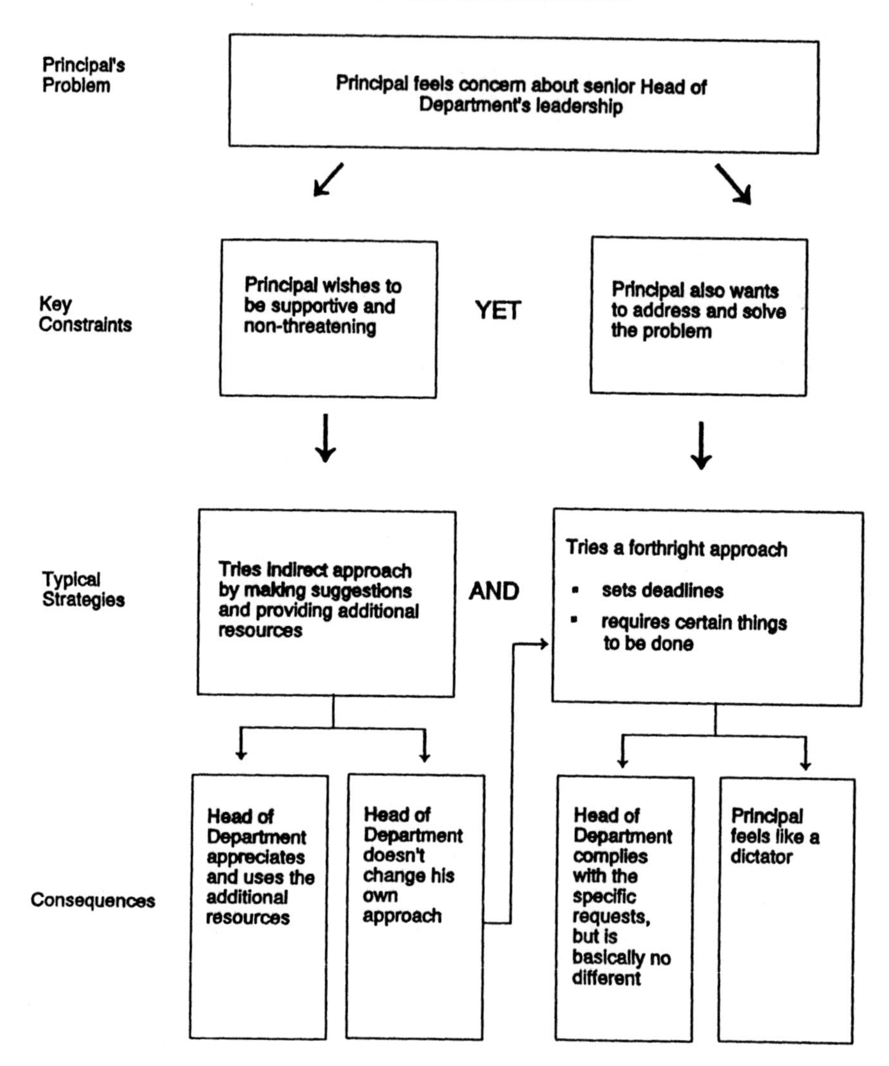

FIGURE 4.2. Analysis of problems associated with a principal's style.

case, despite additional resource allocation, her concern was only superficially resolved, and the deeper disagreements between Carol and the Head of Department remained undiscussable.

In summary, Carol had been unable to resolve the problems in the professional development programme because her view of positiveness worked against the identification, analysis and resolution of problems which required a complete reformulation, rather than the improvement of a few shortcomings. Her decisiveness made it difficult for her or others to acknowledge uncertainty and to allow both the process and outcome of the problem-solving process to be unpredictable. Her unilateral control made it difficult for her to listen deeply to others who had different views about, for example, the worth of job descriptions or of the PDC itself.

In asking Carol to listen deeply to doubts about the impact of the programme and to staff resistance, we were asking her to entertain the possibility that what she thought was positive about the programme was in fact negative, and to face problems to which she had no immediate answers. In short, effective problem-solving, in our terms, required Carol to behave in ways which she herself believed were ineffective. The next step in helping Carol was to discuss and resolve this apparent paradox.

The Adequacy of the Problem Solution

The requirement to develop a professional development programme constitutes an ill-structured problem, in the sense described in Chapter 2. The solution which the programme represents has been judged as inadequate in that it led to the unintended negative consequences which were the trigger to the research contract (Figure 4.1). This solution, which constitutes the theory-in-use of the programme, can be evaluated more formally and comprehensively by applying the criteria of theoretical adequacy described in Chapter 2.

First, the effectiveness of the theory is judged by asking whether or not it produces the intended consequences without violating important constraints. The intended consequences were those of improved teaching and management practice, and these were only achieved when teachers both chose, and were able, to make the connection between their own practice and the development opportunities which they experienced. Occasionally, as in the example of Carol's discussion with the recalcitrant Head of Department, improvements were achieved through non-voluntary processes, but these were temporary, superficial and violated the theory's own value of collegial support. Our analysis suggested that the programme's ineffectiveness was largely due to the fact that staff understood accountability and collegiality in ways that led to an irresolvable dilemma between them. Important problems simply could not be tackled effectively, given that, in practice, collegiality was equated with threat avoidance.

The criterion of explanatory accuracy is not relevant to judging the theory in Figure 4.1, because this theory represents the staff's solution to the problem of how to develop an effective programme, not their theory of why it was ineffective. Explanatory accuracy is relevant to judging the researcher's theory of why this solution was itself problematic, and this will be discussed at length in Chapter 6, as part of a critical discussion of the whole case.

The coherence criterion evaluates how well this theory of professional development, even if it were effective, which it is not, satisfies the constraints set by all our best theory. In other words, an adequate problem solution not only solves the focus problem, but does so in a way that facilitates rather than impedes the resolution of other related problems. This criterion immediately raises the issue of what counts as our best theory. Carol was extremely conscientious about relating her understanding and practice of professional development to the relevant literature, which was her "best theory". The researchers challenged the adequacy of this theory, not only because we believed it to be ineffective, but because it violated other theories about the interpersonal and organisational conditions needed for learning and problem-solving (Argyris & Schön, 1978). In the end, Carol herself used this knowledge to re-evaluate the professional development literature, and thereby revised her "best theory".

The work done by the coherence criterion in this case can be seen more clearly if we imagine that the theory of action portrayed in Figure 4.1 had been effective. I have spoken to some school principals, for example, who believe that it might take five years, but that such a programme will be effective eventually, as their staff become more confident about its benefits and less anxious about its possible threats. Such claims, even if true, invite examination of the way such an approach to professional development would create new problems or exacerbate old ones. The cost–benefit ratio of such a programme would exacerbate the problem of how to efficiently utilise a school's limited resources. A programme which takes five years to make an impact on teacher practice would heighten concerns about the ability of professionals to resolve problems of teacher competence in ways that do not protect teachers at the cost of current students and their parents. The solution suggested by these principals fails the coherence test because it would make it harder to solve other problems associated with the efficient allocation of resources, and with teacher accountability and professionalism.

The improvability criterion examines the extent to which a theory promotes or inhibits detection and correction of errors in its own formulation of a problem. This was the major focus of our examination of Carol's approach to the problems she detected in the programme (Figure 4.2). We concluded that her positiveness, decisiveness and her unilateral control made it difficult for her to recognise both the extent of the problem and the way in which it could not be resolved, given the assumptions that informed its current functioning.

Summary

The professional development programme at Western College lacked impact because it was designed to foster participation and support rather than to critically examine existing and alternative teaching and administrative practices. Although improvement of such practice was a strongly espoused goal of the programme, it was not directly pursued, because doing so would have violated staff's belief that effective professional development was collegial and non-threatening. This avoidance in turn set up a dilemma for those who believed that the programme should be able to address some of the tough issues involved in achieving these goals. The dilemma proved irresolvable because the theories that were brought to its resolution made it impossible for the principal and her executive to simultaneously address these issues and to maintain their conception of collegiality and support. These theories were sustained in turn by the policy context of professional development, the academic literature on professional development that the principal was immersed in, and by the principal's own leadership and problem-solving style. In assessing the adequacy of the school's theory-in-use for professional development, it was argued that it had major shortcomings on the criteria of effectiveness, coherence and improvability.

The story of our subsequent work with Carol and Western College continues in Chapter 7 with an account of the intervention processes designed to help staff address the reasons why the programme lacked impact and to develop a more adequate theory of action to inform both their understanding and their practice of professional development.

References

Argyris, C. (1985). *Strategy, change and defensive routines.* Marshfield, MA: Pitman Publishing.

Argyris, C. & Schön, D. (1978). *Organizational learning: A theory of action perspective.* Reading, MA: Addison Wesley.

Cardno, C. (1988). Problems in promoting the professional development of staff in New Zealand secondary schools. Unpublished master's thesis, Massey University, Palmerston North, New Zealand.

Cardno, C. (1990). Collaborative research: A subject's perspective. Unpublished manuscript.

Codd, J. (1983). The ethics of teacher assessment. *Delta,* **32**, 65–74.

Crane, A. R. (1975). "Supervision". The text of a TV presentation, University of New South Wales, Sydney.

Department of Education (1988). *Tomorrow's schools: The reform of educational administration in New Zealand.* Wellington: Government Printer.

Prebble, T. & Stewart, D. (1983). Professional assessment and the principal. *Delta,* **32**, 45–53.

Stewart, D. & Prebble, T. (1985). *Making it happen: A school development process.* Palmerston North: Dunmore Press.

5

Analysing Educational Problems: The Dilemmas of Participative Management

Of all the educational practitioners I have met, Tony, the principal of Northern Grammar, espoused a philosophy of school management and leadership that came closer than any other to the ideals of critical dialogue. After listening to him describe his ideas and his hopes for the school, my colleague and I wondered what if anything we could offer him. We were confident that he would offer us, at least, a thought-provoking perspective on participative management, and the opportunity to study how such ideas fared in a school whose existing culture was in many ways antithetical to the values and practices which we all espoused. How does a leader who espouses the values of equal worth, participation and internal commitment fare in a school which is hierarchical, and which offers all but its most senior staff limited opportunities for participation? How did Tony understand the transition process from the old to the new culture? What benefits did he anticipate from a more participative management structure, and to what extent were they realised?

This second case enabled us to develop the organisational implications of critical dialogue, which to this point has only been discussed as an interpersonal process. What does it mean to be open and to test ideas with a staff of eighty or ninety, who vary considerably in their experience, expertise and, more importantly, in their educational convictions?

Successful implementation of participative management in a large organisation requires the resolution of several dilemmas, three of which constitute themes in this case. First, participative management with a staff of eighty takes time. How does a leader in a large school foster staff participation, when the extent of staff's involvement is constrained by administrative and teaching deadlines and conflicting demands on time and energy? At what point are the long-term gains that may eventuate from a greater investment in discussion outweighed by the short-term costs that come with postponement of a decision?

Second, how does participative management fit with the widely accepted view (Sergiovanni, 1984) that one of the responsibilities of the school leader is the development and articulation of a coherent vision of the school? A participative management style may make it impossible to meet this expectation, particularly in a school where there is a large and very diverse staff.

Third, how does a leader manage the diversity of views and the inevitable conflict that arises once previously proscribed aspects of organisational functioning are allowed on the agenda? How can the trap of either resorting to autocracy or becoming bogged down in endless debate be avoided? Neither Tony nor we were fully aware of these dilemmas at the outset of our work together. We simply predicted that making the transition would be difficult.

The Research Contract

We all saw the central question as being how to translate the values we shared into the day-to-day reality of school life. Tony was very conscious of the difficulties he would face, and believed that without careful monitoring the pressures of the job would lead him to resort, at times, to autocratic management practices. It was only half in jest, therefore, that he asked for our help in answering the following questions: "Am I an autocrat? If not, will I become one?"

If we were to provide such help, we needed to focus directly on the description and evaluation of Tony's leadership style. Specifically, our research questions were:

1. What is Tony's espoused theory of leadership and school management?
2. What is the theory-in-use that informs Tony's leadership and management practices?
3. How adequate are these theories in terms of Tony's own goals and the researchers' criteria of theoretical adequacy?

The Research Approach

In the previous Western College case, testing the connection between educational problems and theories of action involved a process of reasoning retroductively from the problems of the professional development programme, to the features of the programme's theory-in-use which gave rise to them. Although the iterative nature of problem analysis precludes too sharp a distinction between retroductive and predictive reasoning processes, the latter form featured much more prominently in this case, because the process of inquiry did not start with a problem, in the sense of a practice or outcome which fell short of a desired standard. Instead, the problem was the much more open-ended one of how to implement a vision of participative management in a school which was currently run along quite different lines. Tony's solution to this problem was not yet in place, let alone identified as problematic in any way.

The research questions involved describing and evaluating the way Tony went about solving this problem. The process was predictive in the sense that if the values and strategies he employed in the process were not consistent with those of critical dialogue, then we would predict that his solutions would be judged to be problematic. For example, if he failed to openly disclose and test his views about the implications of participative management, then he should have trouble gaining staffs' commitment to the process of change, and to the resulting decisions.

The analysis presented in this chapter is based on data collected over a five-month period prior to the feedback and intervention phase of the project. During this time we learned about Tony's espoused theory of school management and change from three interviews which were taped and transcribed, and from school policy documents which he had written or which represented his views on school management. Information about Tony's actual leadership style was gathered by observing him in key meetings and either taking extensive notes or taping the meeting. These meetings included six full staff meetings, two Deans' meetings and one meeting of the school's governing body. We also observed Tony give two addresses to outside groups and attended one teacher-only day. One member of the senior management group did not wish the executive meetings to be taped, and our need to respect her wishes limited the quality of our record of Tony's actual leadership practices in this important forum. Information about how others experienced Tony's style was obtained from a series of individual interviews with staff chosen in consultation with Tony and one other informant, to represent a range of reactions to the principal's leadership style and to his vision for the school.

The following account of Tony's espoused theory and theory-in-use has been revised on numerous occasions as a result of discussions between Tony and the two researchers. We should signal, however, that it is a complex construction rather than a straightforward account of the principal's views. Although Tony agrees with its main points, it tends to portray him as much clearer about his views than he felt at the time. The process of constructing his "theory" of leadership is as much one of retrospective attribution of motivation and meaning as of asking him to verbalise what may seem to read below like a "grand plan". The fact that Tony was not fully aware of his theory of leadership at the time, however, does not detract from its importance as both explanation and predictor of his actions.

A New Principal in an Old School

Tony had been appointed principal of Northern Grammar approximately one year prior to the commencement of this study. He came to the position soon after a term as national president of the secondary teachers' union (PPTA). He had had a high public profile during his time in office, and was recognised as a very competent speaker with an assertive and persuasive manner. He was also recognised for his ability to chair difficult meetings productively by succinctly summarising complex issues, and by facilitating the discussion and resolution of complex and controversial debates.

There were several factors that led Tony to sense the importance of leadership style and to invite us to observe and evaluate his practice. First, Tony believed that his notion of school management was substantially different from that of the previous executive team, and that this difference would provide a considerable challenge for the school and for himself. His sense of the challenge involved came, in part, from his experience as PPTA president where he had been aware of schools that were experiencing serious conflict between principals and their staff. He believed that some of these disputes resulted from the autocratic actions of principals, and he was determined to develop both a style of leadership and school structures which spread authority and responsibility more equitably across staff.

Second, Tony had been influenced by a stern warning from another principal that "... (no matter what ideas you have when you begin) after four years you'll become a tyrant—you won't want to but you will". This principal believed that the trend towards tyranny was inevitable as staff withdrew from decision-making and left the principal in a power vacuum. Tony hoped that this project would help him to avoid this outcome. Third, Tony's involvement in this research also reflected his desire to learn some new skills. His route to the principalship was an unusual one, and while he saw many advantages in not having held a deputy principal's position, he also recognised that he had a great deal to learn about the principal's role.

Northern Grammar is an old school and has a sense of tradition and pride in academic, sporting and cultural achievement. It serves a predominantly middle-class European community in Auckland's northern suburbs. At the end of 1987, the staff reflected a range of values along a liberal–conservative dimension, with the executive representing much of this diversity. The principal and senior mistress, for example, were younger, new to the school, and less familiar with and attached to its various grammar school traditions, than were the deputy principal and senior master who were older, and had worked for the school and the community for many years. Tony had sought and supported the appointment of the various executive members, arguing that he needed the diversity of their views in order to be fully briefed about the school and be presented with a range of perspectives.

The Principal's Espoused Theory of School Management

After several interviews we came to understand something of Tony's vision for the school, and his views about how to make it a reality.

Decision-making as Participative

Tony wanted to establish patterns of authority which would ensure that staff and community groups who were to be affected by a decision had an equal opportunity to influence its outcome. Special efforts had to be made to include groups who had previously played little part in the decision-making process. For example, there were very few Maori students at Northern Grammar, yet despite this, he wanted their voice and that of their parents to be heard.

> *Tony*: When it comes down to checking whether what we're doing in this school is meeting the needs of the community, you would say to yourself "has there been any forum in which, say, the Maori community, as big or small as it may be, has had any input? And if there isn't then . . . you go and get some, or make sure somebody is contacted in an appropriate way to do that . . . ."

A Vision for the School

If participative processes were to be purposeful and result in a coherent rather than fragmented school, they needed to be regulated by a common sense of purpose and task. After a year at the school, Tony wrote what came to be called the "Structures Paper", in which he outlined broad objectives for the school and the structure of meetings and responsibilities which he saw as necessary for their achievement. Tony hoped that this philosophy would serve as a touchstone for decision-making so that staff could carry out their roles relatively autonomously, guided by a common philosophy rather than by the decrees of the principal.

> If we want to see this school draw together into a coherent whole we need a touch-stone, a clear philosophy to use in judging every decision as it is made. I want the touch-stone to be this—will this decision help to develop our students as interesting people? [NGDO, 17.9.87].

His paper elaborated what he meant by an interesting person in terms of acceptance of one's own attributes and achievements while constantly striving for excellence. The document concluded with an appeal for the staff to join him in this exciting task:

> I believe the task of unifying the purposes of the school and developing these interesting people will be enormously exciting. When we achieve it, as we will, we will have the best school in the land. I really want that, and I'm prepared to go the extra mile to get it. Especially, I want you all to come with me [NGDO, 17.9.87].

Given the degree of commitment Tony felt to this vision, was he open to having it seriously challenged and even substantially revised during the process of staff participation which he encouraged? His writing in the "Structures Paper" suggests that he was asking others to help him clarify the task and to come on board, rather than to debate the merit of the vision he articulated:

> There are approximately 90 staff at [Northern] Grammar School. About 70 of them teach. It is time for me to establish firmly what my hopes and expectations are.... Over the year I have been waiting for the right moment to make this speech. I feel that now is the right moment. I have already during the holiday asked for and got the commitment of my colleagues in the senior management team to operate according to the philosophy I have outlined.
>
> ... The next step is to win your commitment to the same ends and to continue the clarifying process until it permeates the school. Therefore I will be talking individually with you to see how the areas for which you have responsibility can be developed in harmony with these aims [NGDO, 17.9.87].

Perhaps Tony's vision of a caring school was so widely shared that it would not be seriously challenged. Tony's writing suggests that he did not expect the participative processes he put in place to lead him to rethink his philosophy for the school.

The School Hierarchy as Supporting Staff and Students

Tony's wish to alter traditional power relations among school staff influenced his view about school structures as well as decision-making processes. He visualised the school as an inverted pyramid, with students at the top and each successive layer serving those above it. He wrote:

> Working for and supporting the students are the curriculum teachers, and working for and supporting the students, the curriculum teachers, the PRs [staff with positions of responsibility] and the senior management team is the principal [NGDO, 17.9.87].

This inverted pyramid structure has important implications for the role of the principal. Rather than "lead from the front", Tony saw himself as a chairperson and facilitator who enabled others to get on and do their job in a creative fashion without constant reference to and dependence on him. The principal was an important part of the machinery, but not necessarily the central part.

Conflict Resolution as a Group Responsibility

Tony described some of his views about managing conflict in a paper he wrote for the governing board about what constitutes a healthy school. He identified the recognition of conflict as the first stage in its resolution:

> Amongst other things I identified something which can damage the health of a school—unresolved conflict. I suggested that the first stage in its elimination was a dispassionate recognition of the conflict's existence [NGDO, 10.87].

He saw all parties in a conflict as equally responsible for resolving it, and thought it counterproductive to achieve resolutions by the use of power or manipulation.

> *Tony*: Well, I mean, if I say what is to happen, I just say: "if there's conflict it's got to be resolved . . . you can't proceed with division so a resolution is essential".
>
> . . . the only way to get to it (resolution) is to get an agreed position on all sides.
>
> . . . and that's the beginning of the solution to the problem because it means that neither one group can say . . . "to hell with the (other)". And it means you've got to find a solution . . . [NGTR 2, 12.10.87].

The Principal's Espoused Theory of Change

Given that Tony saw his ideas on school management as very different from what was in place on his arrival at the school, we were interested in his views on how the change process should be managed.

The Principal as a Model

Tony saw it as very important to be able to demonstrate, by his actions, the philosophy of management he was advocating. In the following extract the principal describes how he is attempting to develop a staff consensus on uniform regulations. He wants his views on uniform to be understood as reflecting his wider philosophy of the school as a place which develops interesting students in a caring rather than a controlling atmosphere.

> *Tony*: . . . and so everything that I've been doing up to now about trying to encourage the sorts of things that were in that little document [on school philosophy] and the style of management and the idea that we're there too, working for other people rather than bossing them around. That's all kind of groundwork that's got to be reasonably securely established before you can resolve the uniform issue, and that takes time. I think actually if I wanted to wreck my credibility with the staff I'd walk into a staffroom and say "Right, you know, anybody you see with an earring in, you know, eight detentions." I think that would actually, probably destroy me as a person able to deliver anything in the school [NGTR2, 12.10.87].

The Principal as Deliberate and Careful

Tony saw it as necessary to manage the change in a planned and deliberate fashion to maximise staff understanding and to minimise conflict. New concepts had to be carefully presented, ideas had to be "seeded" and "nurtured", people had to be educated rather than manipulated and the whole process involved careful timing and selection of "the right moment".

> *Tony*: Over the year I have been waiting for the right moment to make this speech [NGDO, 17.9.87].

> *Tony*: That's all kind of groundwork that's got to be reasonably securely established before you can resolve the [uniform] issue and that takes time. . . .

> I have taken some care to see that the two sets of opinions are going to be expressed at about the same time in the year, so that they can in fact be seen to be different. Now that enabled me to kind of seed in an approach into the debate . . . [NGTR2, 12.10.87].

Tony's patience in waiting for the right moment could be inconsistent with his earlier claim that conflict should be faced up to and resolved by all involved. His efforts to introduce ideas about change to people in carefully managed gentle ways could be designed to avoid difficult discussions. When challenged by us on this possible inconsistency, he clarified his own approach to conflict this way:

> *Tony*: No, but I definitely don't seek it (conflict). And I will go to great lengths to avoid it. No, that's not the right word. I'll go to great lengths to say what needs to be said, in ways that are not conflicting. Except when I lose my cool, which I try not to do [NGTR6, 3.11.87].

Tony believed that his ideas were not inconsistent because his efforts to prevent conflict were not incompatible with facing up to it when it did occur.

Evaluation of the Principal's Espoused Theories

There is a broad equivalence in tone and intent between Tony's vision for the school and the values and strategies of critical dialogue. He was striving for a supportive, co-operative environment where all felt cared for and able to care for others in the process of developing interested and interesting students. Activities of school members would be regulated by a shared set of principles rather than by the traditional authority of the principal, and all would have the opportunity to influence school decisions.

Although Tony did not comment on it directly, this view of participative management can address some of the problems of inefficiency commonly ascribed to it. As staff work through specific decisions, they should gain a clearer sense of what their shared vision means in practice, and hence of which options violate and which are consistent with its principles. As school policy decisions become more widely shared and more widely understood, the complexity and number of debates should reduce.

The means Tony articulated for achieving his vision were less congruent with critical dialogue than the ends he hoped for. He saw his task as "winning staff over" rather than debating with them the merits of the vision he articulated. The purpose of critical dialogue is not to persuade people of one's ideas but to test those ideas and to develop commitment to whatever ideas survive this process.

Tony anticipated that in a staff that was roughly divided along liberal–conservative lines, there would be opposition to the implementation of some of his ideas. He talked, therefore, of the need to manage change carefully—to educate rather than manipulate, to seed and nurture ideas. The line between the manipulation and the education of others is a very fine one. One point of difference between them is whether the strategies employed are kept secret; another is whether the educative processes are reciprocally or unilaterally intended. Thoughtful planning about how to introduce change is not incompatible with being democratic. Planning that remains private, is not checked and is designed to maximise the chance of influencing others while remaining uninfluenced is manipulative and incompatible with critical dialogue. In an effort to avoid upset to others and to seek a productive debate, Tony made plans that made a series of untested assumptions about the pace of change that was acceptable to others, about how to handle perceived or anticipated conflict, and about the sequence in which to introduce the various steps. His writing also suggests that he saw the conversations with his senior staff as opportunities to win them over rather than to test the merits of his ideas. In these ways, his understanding of the change process was a manipulative one, with much in common with the unilaterally protective and controlling features of what Argyris calls Model One.

We were now in a position to answer Tony's question about whether or not he was autocratic at the level of his ideas on school management. His description of how to manage change was manipulative and autocratic, despite the vision itself being participative and democratic. His ideas on change were a response to the dilemma he felt between his wish to implement his vision and to minimise the conflict that it might engender with those staff who preferred the old management style and structures.

The Principal's Theory-in-use for School Management

Figure 5.1 describes Tony's theory-in-use for solving the problem of how to introduce participative management into a hierarchical school. The contradictory constraint structure that he set to solve this problem placed him in two different dilemmas. First, he was caught between his desire to foster participative practices and his desire to win over his staff to his vision of the school. Second, given that his vision was in conflict with the existing school culture, its implementation would, at times, jeopardise his desire to minimise conflict and emotional unpleasantness. The result of these tensions was a set of strategies that left staff appreciative of the participation opportunities, but confused about their purpose, and uncertain of their own and their principal's commitment to the change process. The evidence for each of these features of Tony's theory-in-use is described in the following section.

Encourage Participation

There was considerable evidence that Tony fostered the involvement of his staff in the ways he described to us. At the structural level, his staff

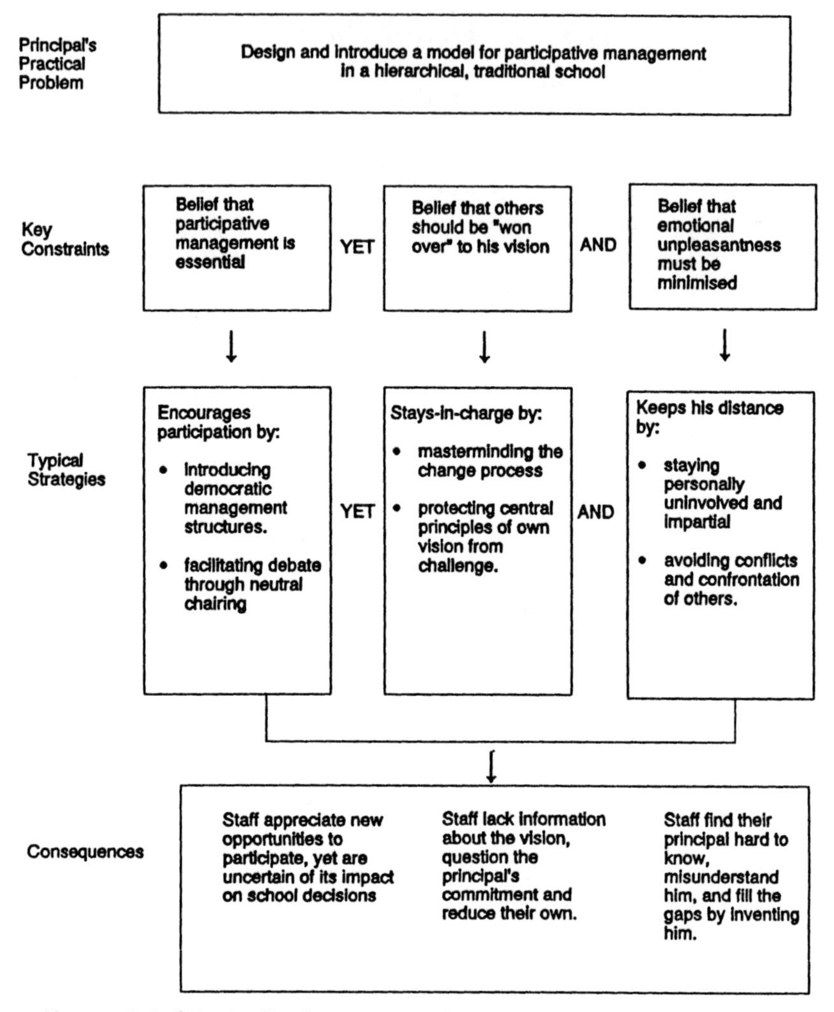

FIGURE 5.1. Principal's theory-in-use for the transition to participative management.

reported that he increased the profile of Maori students by appointing Maori teachers and introducing Maori protocol in school ceremonies [NGTR 14, 19.11.87]. Similarly, they reported he implemented union policy on the role of women in schools by making or recommending more senior women appointments [NGTR 14, 19.11.87]. His insistence on a team approach to school management at the executive level was also an attempt to share power and thereby protect the school from the possible excesses of any one person [NGTR 28, 15.2.88].

At the process level, Tony went to considerable lengths to encourage greater staff participation, particularly of those groups such as women and junior staff who were reluctant to speak up. At staff meetings, for example, he would introduce a topic for discussion without describing his own views, and then control the speaking turns of the staff so that participation was balanced according to the gender and seniority of staff. Frequently, no conclusion was reached; the purpose for Tony was that staff had had a chance to speak and to hear the diversity of views of their colleagues. By and large, staff experienced this meeting style positively.

> *VR*: How do you see that (decision-making) process . . .?

> *Staff*: Well, I see that process as being certainly more open than it used to be—there is this discussion process we go through now [NGTR14, 19.11.87].

The invitation to staff to contribute to meeting agendas was also seen as a sign of greater openness.

> *Staff*: . . . it's something that is just starting to change its nature, in that issues are being discussed, people are being asked, "Is there anything you want to add to the agenda?" [NGTR4, 20.10.87].

STAY-IN-CHARGE

In contrast to these efforts to share power and opportunities for influence, we also detected patterns that indicated that Tony wanted the participation of others to serve predetermined ends. His ideas about careful planning and strategising were at times reflected in his practice. He released documents when he thought the school was ready and waited for the right moment to approach people on what he predicted would be a difficult issue. Tony's delays were frequently motivated by a wish to avoid unnecessary stress or conflict, but he unilaterally decided how to judge others' stress and how to manage it. An executive staff member comments on his behind the scenes planning of group tasks:

> *Staff*: If you look at how these things are allocated, I think he has got an idea in the back of his mind as to who would work together with whom to the best advantage [NGTR13, 19.11.87].

During his term as president of the PPTA, Tony had learned a set of skills that enabled him to adopt a professional role rather than a more emotional and personal approach to his work. By adopting such a role he kept a reflective distance from the competing demands made on him, and from the reactions of others. By training himself not to react emotionally, Tony could survive in situations which others found extremely stressful. For example, part of a teacher-only day held in 1987 involved teachers working in small groups to identify those things that contributed to stress in their jobs. Two senior staff members were very upset by what they had perceived as the negative and blaming tone of the subsequent feedback session. Tony, on the other hand, had not taken the feedback personally, and was pleased that the staff had felt able to speak so freely.

At other times, Tony's reflective style delayed rather than eliminated his emotional reactions. After our first feedback sessions with him, Tony phoned us with his reactions:

> *Tony*: I thought I should tell you how I had analysed my thoughts after I'd had time to go away and draw breath . . . which I find is a way I do react. I tend not to react straight away to events that happen. I think that's why I used to survive negotiations in [the PPTA] . . . I was able to remain impassive, under attack from the [Minister] . . . or whoever it happened to be, and then I fell apart in the evening. . . . And the same thing happened again. So that in a sense when we were sitting here, you asked me how did I feel, I really couldn't answer because I didn't feel anything [NGTR30, 3.3.88].

Another feature of this stay-in-charge approach was his unchecked positive evaluation of school activities and processes. In the final session of the 1987 school review day, for example, Tony and the executive listened without comment to the concerns of staff about the organisation of the school and the functioning of the executive team. Tony then amazed some of the staff by referring in his concluding comments to a list on the staff room notice board, which in his view meant that nearly all their concerns were currently being addressed. We use this example as evidence of his "stay-in-charge" strategy, not because his evaluation differed from that of many staff, but because he failed to check his evaluation that such a list indicated "progress" [NGTR19, 27.11.87].

Keep Your Distance

Tony anticipated that the changes he planned would alienate some groups of staff, making it difficult to forge a consensus between the more conservative and more liberal groups on the staff. He thought he would be more effective in these circumstances by "keeping his distance", which involved avoiding becoming too closely identified with any one staff group. In addition, Tony was new to the job, and by staying impartial he gave himself time, as he put it, "to sort out the try-ons". Tony also "kept his distance" by avoiding becoming emotionally involved in the politics and personalities of his job.

Tony's wish to separate out his professional role from his more personal actions and reactions affected our ability to uncover the reasoning processes and values that lay behind his use of particular strategies. We wanted to describe not only the typical strategies he employed, but also why he used them, and this required discovering some quite personal information. We learned early on that it would be difficult for us to get to know Tony in this way.

> *Tony*: You're saying you want to get behind my skin.
>
> *VR*: Yes.
>
> *Tony*: Well I'm not going to let you. But that's just my personal thing. I do have my barriers up, I've done it all my life . . . and in a way, I would say, I will say as you want to know what I feel, that it has been a cause of my ability to do what I have done in the past well, has been the fact that I have never been cliquey or factional in any work place that I have ever been in (. . .) and I know when I was in the PPTA—this was how I was successful in negotiations where others were not, because I was in fact able to relate to those people in their role—my role to their role, I was able to do that. That's because I am happy to go in there living a role. I can keep the personality out of it [NGTR25, 4.2.88].

In this extract Tony described his adoption of a role as a habit which he used easily and naturally, and which enabled him to be successful where others had failed. Several staff commented on the difficulty of getting to know their new principal and one attributed this directly to his wish to separate the role of the principal from the rest of his persona.

> *Staff*: [Tony] didn't really think anybody knew him. . . . I mean there are messages in that.
>
> *MA*: What were they?
>
> *Staff*: Well, the kind of relationship that I had had with people I'd worked with in the past was obviously not what he felt comfortable with. . . . I mean he was going to be in the role of his job, and there was a certain tiny amount of himself that he was willing to share, but the rest was strictly private [NGTR26, 8.2.88].

Tony's strategies appear to serve an interesting mix of values, with "keep your distance" and "stay-in-charge" serving values of unilateral protection and unilateral control respectively. The third strategy of "encourage participation", however, is much more difficult to evaluate because participative styles of leadership can serve either manipulative or more democratic goals. We believe that both values were probably operating for Tony. At times, he fostered genuine debate and was open to reciprocal influence, and at others, participative management was linked to his "stay-in-charge" and "keep-your-distance" strategies and served manipulative ends. This ambivalence probably reflects Tony's wish to foster participation while protecting his vision from substantial challenge. At times he was open, at others, for example over corporal punishment, too much was at stake to allow conservative opinions to jeopardise his plans.

Consequences of the Principal's Theory-in-use

Approximately one year after Tony's arrival at the school, we asked a range of staff for their views on his management style. From these reports and our own observations, we drew some tentative conclusions about the consequences of his attempts to unite the school around a shared philosophy by encouraging staff to participate in decision-making, while he himself sought to keep his distance from the inevitable political and personality conflicts that such processes would engender.

Appreciation of Participation

There was general recognition of, and appreciation for, the way Tony involved more people in the management of the school. For some staff his approach had led to a feeling of greater involvement in the school:

> *Staff*: . . . well, I think that Tony has brought a level of leadership which is consultative and fair . . . [NGTR10, 12.11.87].

> *Admin*: I think so far as the admin area, we are thought more of now as staff members. . . . He's aware that everybody in the school, down to the tuck shop workers, the assistants and the ground staff, are people [NGTR27, 15.2.88].

People who had never spoken previously at a staff meeting expressed their views publicly for the first time as they sensed that a more open decision-making process was operating.

Doubts About the Purpose of Participation

While many staff appreciated the increased participation, they also questioned the connection between the process of staff discussion and the decisions that were finally made. Decisions appeared "magically out of nowhere" or came into being without staff being properly informed about what they were or how they had been reached.

> *Staff*: . . . after much discussion where there has been no concrete conclusion to me, at times decisions seem to appear magically from out of nowhere and no-one is quite sure who made it, or

where it came from ... that sometimes leads to ill feeling ... [NGTR14, 19.11.87].

Staff: Yes, but he doesn't actually always say, this is [the decision].

VR: How does he put it?

Staff: He may not put it, it comes into being and I think that is part of the management style, which can tread on some people's toes [NGTR12, 19.11.87].

Given that the principal wished to enable staff to accept responsibility for jointly running the school and to alter school structures and meeting processes to facilitate this, how is it possible that staff believe that this sharing often ceases when the final decisions are made? Tony's beliefs about change, described in the previous section of this chapter, certainly suggested that he did not see the basic principles of his vision for the school as being negotiable. Given this, one might predict that decisions made would not always reflect the debates in which staff had participated. Surprisingly, perhaps, there is little evidence to suggest that staff felt the decisions taken were contrary to staff discussion. Their concern was more that the process was inefficient and badly organised, resulting in premature foreclosure of the decision process when time ran out.

A possible explanation for the apparent disconnection between staff discussion and the eventual outcome relates to Tony's reflective style. His tendency to manage the sequence and timing of others' involvement, and to avoid announcements which could upset others, would have left staff without a shared understanding of the process by which they were to arrive at decisions, and of the progress that various groups were making along the way. It is inevitable that a principal who wishes to introduce a process that is new to staff has a greater understanding of its purposes than others. Leaders can reduce misunderstandings, however, by frequent checking of the match between their own and others' understandings of the process in which they are engaged. Such checks would have provided Tony with an early warning of staff's concerns about the value of the discussion process. While Tony continued to assume responsibility, however, for the private "behind the scenes" management of those processes, he was unlikely to use these processes of public checking and reflection.

Staff "Invent" their Principal

A constant theme of our early interviews with staff was their difficulty in getting to know their principal. They knew the broad outlines of some of his educational ideas, for they had read about them in a memo to staff and in his contributions to the national *Curriculum Review* (Department of Education, 1987). This left them dissatisfied, however, because the translation of those ideas into the complex reality of Northern Grammar was far from obvious. They wanted to know more about their principal's thinking on these matters, not only to be sure that he was working with them on those complexities, but to be convinced that he was as committed to the concrete practice of the ideas as he was to the ideas themselves.

Aspects of Tony's management style made it very hard for him to meet these expectations. He could not be a full participant in the debates about how to practice the educational principles he espoused, while at the same time acting as impartial facilitator of the debate. However, by not fully participating, staff began to question his commitment to those ideas. The problem was compounded by Tony's very private approach to much of his job, which left staff with very little idea of the sorts of dilemmas that he was experiencing in coming to grips with the role of principal.

In the absence of first-hand information about Tony's ideas and feelings, and ignorant of the reasons why they knew so little, staff began to fill the vacuum by inventing aspects of their principal. They projected onto Tony their own ideas about his motives, values and commitment. Everything from the reason why he looked out of the window when he talked to them in his office, to why he left school early at times, to why he continued his connections with the media and the national teachers' union became the subject of speculation.

> *Staff*: ... Does he really care about people? The articulated philosophy may or may not be what he really believes ... you can be in the middle of a sentence and he will look away and absolutely cut you off ... so that you get the feeling that he is never actually there. He's never actually engaged [NGTR22, 17.2.87].

> *Staff*: It's a thing you get, I guess, if I feel I don't know the person. I mean that's the cost to him as well, with the rest of the staff. The review day goes back to that thing. If you don't think you really know the person, then unless you're going to reveal even a wee bit of you to me, how am I going to establish trust? [NGTR26, 8.2.88].

When Tony read the first draft of this chapter, he disagreed with about a third of the comments of staff which were included in the text. In his view, their beliefs about him were either wrong or oversimplified. He wrote this comment, for example, about the quote in which he is described as not being engaged and "cutting off" the conversation.

> *Tony*: People demand "me" without any regard for whether "I'm" available. In no way a cut off—just thinking through.

The staff member had interpreted Tony's pattern of non-verbal behaviour as a "cut-off" and built this "data" into a hypothesis that the principal was not really committed to his espousal of a caring philosophy. Tony, on the other hand, knew his own behaviour was part of a characteristic pattern of reflection; a pattern which he believed helped him to be more effective.

Why did these misunderstandings of the principal develop and persist? The reason lay in the interaction between Tony's actual management strategies and the reasoning of his staff. Tony's strategy, which we called "keep your distance", left staff short of critical information about his motives, intentions and thought processes. While this strategy helped him to avoid over-identification with a particular staff group, the price he paid was staff misunderstanding of his style and of the dilemmas he experienced. The staff observed his performance of the role, but were left to invent the reasoning processes behind it.

Some staff contributed to the development and persistence of these misunderstandings through their willingness to draw conclusions about Tony on the basis of unchecked speculation about his intentions and motives. Once drawn, these conclusions became the taken-for-granted evidence upon which further conclusions were drawn, so that in the end, these staff became convinced of Tony's lack of commitment to his stated philosophies. To be fair to the staff, however, we need to consider this sequence from their point of view. What would have been involved in checking some of their speculations? It would have involved a conversation which required them to speak quite personally to a man whom they believed did not welcome that type of relationship. Some staff had attempted such conversations in the past and felt dissatisfied with the result because they did not trust the answers they received, or because they felt disadvantaged by the considerable debating skill of their principal. In addition, Tony may not have been willing to provide the sort of personal information they wanted because he may not have recognised the emotional and personal nature of their request, nor have been able to reciprocate in kind.

Evaluation of the Principal's Theory-in-use

How adequate is Tony's solution to the problem of how to introduce a more participative management style to a school which was run on more traditional hierarchical lines?

On the criterion of effectiveness, the evidence suggests that the theory (Figure 5.1) represents an only partially effective solution. Tony's genuine commitment to participative management led him to restructure senior roles and responsibilities in ways that empowered junior staff, and to massively increase the opportunity for all staff to participate in decision-making. The theory was far less effective, however, in fostering the shared commitment and understanding that Tony had so passionately asked for in his "Structures Paper". Tony attempted to resolve the dilemma between being committed to his vision and imposing it on his staff by increasing staff's opportunity to have their say, while simultaneously limiting his own role in these debates and in the day-to-day practicalities of implementation. While this reduced the risk that he would fall into the trap of autocratic leadership, the price was a perceived incongruence between his espousal of his vision and the impartial facilitation that he exercised in many meetings. As a result, staff were short of information about how Tony understood the practicalities of his vision, and they began to mistrust his level of commitment to the implementation process. Both consequences reduced the probability that staff and principal would gain a sufficiently deep and shared understanding of the vision to use it as a guide to decision-making in the way that Tony espoused.

The coherence of Tony's theory can be questioned at the level of its internal and external consistency. At the internal level, Figure 5.1 reveals tensions if not contradictions within the theory. One cannot be both committed to participative management and protect one's vision from challenge through masterminding the change process. If the leader is not open to testing central features of his or her vision, the parameters of participation are repeatedly constrained by its non-confrontable elements. In addition, there was for many staff an incongruence between the passion with which Tony espoused his vision and the low-key approach he took to working through the complexities of its implementation.

At the external level, the theory which is summarised in Figure 5.1 raised problems related to the efficiency of the various debates and the time that was taken from other essential staff roles. Tony's impartial leadership in meetings left staff frequently unclear about the decisions that had been made and the next steps that would be taken. At the same time, some of the teaching staff began to question the appropriateness of a management structure that left them feeling that they shouldered as much of the burden of school administration as did those with specialised administrative roles. A theory of participative management must cohere with what we know about organisational differentiation (Mintzberg, 1979) and differential reward structures.

The improvability criterion was valued by Tony because, in inviting the researchers to monitor his leadership, he had acknowledged that there were many skills he needed to learn. In addition, he was insistent that he was feeling his way in terms of the details of the change process he was leading. Despite his espousal of learning, his theory-in-use contained features which prevented him from doing so. These features are, on the whole, related to the high value he placed on unilateral protection of self and others from unpleasantness. Given this value, he could not publicly test his ideas and perceptions with those he believed might become upset by them, and these individuals were the very ones whose competing theories were likely to generate the strongest objections. As a result, differences between individuals remained undiscussable, and opportunities for both parties to learn more about the implications of their theories were lost.

Summary

Tony's vision for the school was a sophisticated version of participatory democracy linked to the achievement of liberal educational objectives similar to those articulated in New Zealand policy documents in the late 1980s. When it came to implementation of this vision, however, Tony was caught in the same dilemmas as many liberal reformers. The first of these dilemmas sets up a tension between the leader's role as developer of a vision for the organisation and the participative principle of providing equal opportunity for all to develop such a vision. The second is the potential conflict between efficiency and democracy (Rivzi, 1990; Walker, 1990), and the third involves the management of diversity and conflict in ways that neither suppress them, nor allow them to paralyse the functioning of the organisation.

Tony attempted to resolve these dilemmas by employing the strategies that had worked successfully for him in the past. By massively increasing the opportunity for staff to participate in decision-making and simultaneously avoiding leading key discussions, Tony hoped to equalise power relationships in the school and thus avoid the trap of autocratic leadership. At the same time, however, aspects of Tony's management style, namely his "stay-in-charge" and "keep-your-distance" strategies, led to an information and credibility gap which staff came to fill through speculation about their principal's values and motives. Tony's style made it difficult for him and his staff to detect and correct these misunderstandings, and the end result was a considerable discrepancy between the interpretations of some staff and the principal of the latter's actions.

This discrepancy was serious because it jeopardised the improvement and the implementation of his vision. The discrepant perceptions of Tony and his staff created a climate of misunderstanding and mistrust, which in turn led to a decreased commitment to the ideas which they had initially endorsed. The researchers believed that the skills of critical dialogue would provide Tony with more effective ways of advocating for his vision while responding openly and sensitively to the emotional and intellectual challenges involved in its debate and implementation. Our attempts to help him in this way are described and evaluated in Chapters 7 and 8.

References

Department of Education (1987). *The Curriculum Review: Report of the committee to review the curriculum for schools.* Wellington: Government Printer.

Mintzberg, H. (1979). *The structuring of organizations.* Englewood Cliffs, NJ: Prentice Hall.

Rivzi, F. (1990). Horizontal accountability. In J. Chapman (Ed.), *School-based decision-making and management*, pp. 299–324. London: Falmer Press.

Sergiovanni, T. (1984). Leadership and excellence in schooling. *Educational Leadership*, **39**, 5.

Walker, J. C. (1990). Functional decentralization and democratic control. In J. Chapman (Ed.), *School-based decision-making and management*, pp. 83–100. London: Falmer Press.

6

Reflections on Problem Analysis

A case report presents the results of a researcher's best thinking, and so inevitably omits much of the reasoning that led to the conclusions which are presented. One consequence of this reporting convention is that the dozens of theoretically informed choices that are an essential part of the data collection, analysis and reporting processes are hidden from the critical view of the reader; a second is that the research process is portrayed as misleadingly linear and straightforward, rather than as iterative and error prone (Ball, 1990; Van Maanen, 1988).

Ball suggests that critical reflection upon these choices is not only necessary to fully inform readers about qualitative research processes, but that it is a requirement of methodological rigour. Readers of quantitative research expect full reporting of data collection instruments, and since the equivalent instrument in qualitative research is the researcher, so should they expect an accounting of the reasoning that informs critical decisions in qualitative research.

> It is the requirement for methodological rigour that every ethnography be accompanied by a research biography, that is, a reflexive account of the conduct of the research which, by drawing on field notes and reflections, recounts the processes, problems, choices, and errors which describe the fieldwork upon which the substantive account is based (Ball, 1990, p. 170).

This chapter serves some of the purposes of Ball's researcher's biography, by critically reflecting upon the processes and outcomes of the problem analysis phase of the two case studies. One purpose of these reflections is to begin to judge the theoretical adequacy of the problem analyses presented in Chapters 4 and 5. This involves describing the steps that the researchers took, or should have taken, to improve the explanatory accuracy of these theories and evaluating the final result. A second purpose is to compare selected aspects of data collection and data analysis procedures in PBM with those found in other forms of qualitative inquiry. This comparison is not intended to be a comprehensive description of PBM procedures, but to highlight those aspects which differ from other types of qualitative practice.

Improving Explanatory Accuracy

In PBM, the explanation of educational problems is guided by the general hypothesis that the theories of action of relevant practitioners are causally implicated in the practices that are called problematic. This hypothesis directs the explanatory search to the understandings and actions of individuals, and to the way they in turn sustain and are sustained by the cultural and organisational practices in which they are embedded. The nature of the understandings and actions, and the way they interrelate to produce the problem, are discovered in the course of theory development for each particular problem. Given this overall explanatory framework, two types of question could be raised about the explanatory accuracy of each case. The first more general question is whether the theory of action framework provides an appropriate causal account of these problems; the second more specific question concerns the accuracy of the way this framework is employed. The following sections discuss some of the steps taken to improve the accuracy of our explanatory theories.

Error Detection and Correction

Defence of the accuracy of an explanatory account requires, in part, a description of the steps taken to detect and correct errors in the emerging theory (Phillips, 1987, p. 22). Quantitative research has well-established procedures for ensuring that disconfirming evidence is noticed and given appropriate weighting in the data analysis process. Qualitative research lacks such procedures, but the logic of seeking to test hypotheses through a process of disconfirmation applies equally to both modes of inquiry. Cronbach (1980) describes this approach to validation as follows: "The job of validation is not to support an interpretation, but to find out what might be wrong with it. A proposition deserves some degree of trust only when it has survived serious attempts to falsify it" (p. 103).

The validity of qualitative research, including the analyses presented in the previous two chapters, is enhanced by showing how it has been modified as a result of exposure to opportunities for disconfirmation. For example, the hypothesised connection between theories of action and educational problems can be tested by raising alternative explanations which do not implicate practitioners' understandings and actions. It is possible that the principal and staff in the first case had a perfectly adequate understanding of professional development and of how to achieve it, and that the limitations of the programme were due to lack of resources. While this alternative explanation was not explicitly considered in the case report, and perhaps should have been, there is sufficient relevant data presented to show its implausibility. First, an under-resourced programme would not have been held up as a model by other schools. Second, the principal had identified the problems precisely because she judged the benefits of the programme to be disproportionate to the considerable resources allocated to it. Third, our investigation of the processes employed in the programme suggested that no matter how generous the resources, the desired impact would not be achieved, because staff were reluctant to publicly examine the tough issues associated with improving the quality of teaching and administrative practice.

The detection and correction of error involves attempts to disconfirm those hypotheses currently in favour, as well as a search for plausible alternatives. An early hypothesis about Carol's leadership style represents an example of the former. We originally attributed to her an unwillingness to encourage or engage with the challenges of her colleagues, but as a result of a deliberate search for counter-examples we realised that this hypothesis was over-simplified. We could recall and locate in the transcripts examples of where she had reacted openly to challenges from a Head of Department whom she respected, and from one of her executive team. The result of this search was a more complex and more accurate formulation of the conditions under which Carol was more and less likely to be open.

Member Checks

Member checks is the term given by some qualitative researchers to the process of negotiating research accounts with participants (Lincoln & Guba, 1985; Van Maanen, 1982). For some researchers, such as Lincoln and Guba (1985, p. 314), this is the primary method for establishing credibility in qualitative research. The emphasis in PBM on member checks as a process of theory improvement is somewhat different from that of these authors. There is little recognition by Lincoln and Guba that the process of gaining consensus *per se* may not, depending on the critical acumen of those involved, bear any relationship to the accuracy of their perceptions. As Phillips has suggested, the history of science shows that it is possible to establish a community of adherents for most beliefs, regardless of their warrantability (1987, pp. 89–90). Gaining agreement about a problem analysis is neither a necessary nor a sufficient condition for establishing its explanatory accuracy. Such agreement increases the probability of accuracy if it is gained under particular conditions, and it is these conditions which are highlighted in the application of member checks in PBM. First, the purpose of such checks must be the improvement of theory through the detection and correction of error, and not the gaining of agreement *per se*. Second, the means to such improvement must be examination of the warrant for particular beliefs, and this implies engagement in a critical dialogue as described in Chapter 3.

In PBM research, participants are involved in member checks for both scientific and ethical reasons. Regarding the former, participants can use their extensive experience of the subject matter to raise objections to researchers' analyses. The result, one hopes, is a theory which is more accurate and more effective because it has been revised to take account of such challenges. When the subject matter comprises attributions about the members' understandings, motives and desires, such checks seem particularly apposite, since actors have a privileged, though not necessarily infallible, access to this type of evidence. One source of fallibility is people's blindness to possible incongruence between their espoused theory, which is typically the source of their judgements about themselves, and their theory-in-use. If the researcher's attributions about intent, for example, are based on the latter, and this is incongruent with what the actor espouses, then disagreement between the two parties is inevitable. Resolving this disagreement through critical dialogue is highly interventionist, since it involves a shift in self-perception, and a possible re-evaluation on the part of the actor, of the theory that was the subject of the disagreement. Examples of the process of resolving these disagreements are presented in the subsequent two chapters on the intervention phase of the case studies. From the ethical perspective, member checks are important, because researchers who base critical evaluations on the basis of unchecked attributions unilaterally control their research participants. In addition, an intervention cannot proceed collaboratively while unacknowledged disagreements about the theory of the problem remain.

It is important in discussing member checks to be clear about what it is that is being checked, because different procedures are required for different types of evidence and analysis. At the level of observed phenomena, the researcher needs to check that he or she is working with data that are reliable; that is that relevant parties agree that his or her records accurately capture what occurred or was said (Le Compte & Goetz, 1982, p. 41). In our cases, this level of agreement was gained by tape-recording and transcribing meetings, and by reviewing interview notes with interviewees, either at the end of an interview or at a later date when clarification was required.

Another level of checking involves the accuracy of the researcher's reconstruction of the theory or theories of action relevant to the focus problem. Since these are made up of the way that actors themselves engage with and understand their world, member checks, as already discussed, are a particularly appropriate way to improve their accuracy. Participants are unlikely to find a critique of their theory of action credible if they believe that that theory has itself been misunderstood. We were constantly amazed in our discussions with Tony and Carol at the scope for misunderstanding that this process held. At times we mistakenly believed that our language was shared; at others, we overestimated Carol's and Tony's ability to articulate the values and understandings that informed their practice. An example of the former occurred when, after several workshops on the theory and practice of critical dialogue, we heard Carol describe it to a colleague as "manipulative". Our dismay was based on the assumption that we shared an understanding and a negative evaluation of that term. A check revealed that she was using manipulative in its literal sense of "like a tool" rather than in its pejorative sense of giving a false impression. Without this check, we could have misjudged Carol's understanding and the progress of our intervention.

A third type of check involves negotiation of the accuracy of the researcher's critique of the practitioner's theory. Some authors argue that this type of member check is inappropriate if the researcher's theory is drawn from a highly specialised knowledge base which is unlikely to connect with the experience and understandings of those who are the subject of the critique (Phillips, 1987, p. 20). The relevant community in such cases is thought to be the researcher's peers who will employ the standards of their discipline in judging the analysis. While not denying the relevance of the latter, participants must also be included in this type of check in PBM, because the purpose of the researcher's critique is to show how practitioners' current understandings and actions thwart them in the achievement of their own goals, or in the fulfilment of their own values. If this critique is not accessible to relevant practitioners, it will not provide them with a compelling reason to alter their practice. One of the purposes of this check is to develop an outsider critique which eventually becomes part of the insider's understanding (Van Maanen, Dabbs & Faulkner, 1982, pp. 17–19). A successful critique changes the boundaries, in other words, between what is "inside" and "outside", and this will not happen if it is not understood or accepted by practitioners.

The negotiation of our critique made a considerable difference to our analyses of the two cases. At Western College, our initial analyses of the problems of the professional development programme, and of Carol's approach to solving them, had failed to recognise the external influences which had shaped her solution to the problem of how to design the programme. As a result, although she agreed with our analysis as far as it went, she rightly felt that the omission of these influences resulted in an overly critical and blaming tone.

While negotiation with Carol led to an enrichment of our analysis at Western College, our discussions with Tony led to even more substantial changes. Our interviews with staff had yielded numerous negative comments about the congruence of Tony's practice and pronouncements, and about the success of the new management structures. Given the consistency and detail of these reports, we had accepted them, in our early drafts, as evidence of the practice of Tony and his executive team. After Tony's denial of the accuracy of either staff's observations, or, more commonly, of their interpretations of those observations, we began to reframe the problem as the consistent disparity between Tony's perceptions and those of many of his staff. The theoretical task was now to explain why this discrepancy had arisen and persisted. In addition, we began to discuss with Tony its implications for his desire that the staff work as one towards a shared vision for the school.

It may already be clear to the reader that our "member check" procedures involved hours of discussion with the research participants. Its purpose was not to sanction our analysis, or to gain participant's co-operation—it was to thrash out theoretical differences as a prior step to beginning an intervention process. As will become obvious in the next chapter on intervention, the struggle towards mutual understanding continued throughout the subsequent intervention phase, as did the process of theory improvement.

Judging Explanatory Accuracy

A collaborative mutually educative research process should apply the same criteria of theoretical adequacy to both practitioners' and researchers' theories of a problem. The relevant criteria were applied to the practitioners' theories at the end of each case; it is time now to turn to the evaluation of the researchers' theories of the problem. In Carol's case this involves evaluating our theory of why the school's professional development had less than the expected impact on teaching and management practices, and why a few staff remained resistant to its provisions. In Tony's case this involves evaluating our theory about why Tony did not achieve his vision of a participatively managed school to which all were committed. This section addresses the explanatory accuracy of our theories of these problems. In Chapter 8, after the intervention process has been described, further questions are asked about the remaining criteria of theoretical adequacy.

In the Western College case, the theory summarised in Figure 4.1 was the result of discussion and revision between the two researchers, and those directly responsible for the professional development programme. These people, especially the principal, had every reason to challenge the researchers' analysis, because by accepting it they confirmed their own role in the problems. In fact, the researchers were criticised by one executive member for using the programme as the context for analysing Carol's leadership style, because her heavy investment in this area meant that, in his eyes, it was here that she had the most to lose from any possible criticism. The principal and executive did raise questions which led to revisions, and their acceptance of the "final" version suggests that they believed it to be an accurate and useful analysis of their difficulties.

Despite the deliberate search for disconfirmation by the researchers and the lengthy member checks, the analysis is still criticisable on a number of grounds. Some simple quantitative data could have provided a better test of our claims about the relationship between threat avoidance and low impact. First, we could have documented the percentage of scheduled PDC consultations that were actually carried out by the executive staff, and the percentage of the recipients of these consultations who had instituted the process with their own staff. Secondly, we could have surveyed the whole staff, and asked them how threatening they personally found the idea of being appraised, and how important they believed that threat avoidance was to the success of the process. This data would have enabled us to document more precisely the staff culture in which Carol was working, and to examine more closely its impact on her leadership style. As it is, our description of the leader lacks a clear sense of the influence of the followers, and of their likely impact on the changes that our theory implied were needed to increase the programme's effectiveness.

In the Northern Grammar case, our analysis showed that Tony attempted to achieve his vision of the school in ways that were counterproductive for achieving his goals of teamwork and shared commitment to a participatively managed school. The theory which is summarised in Table 5.1 is the result of extensive consultations with Tony, and numerous revisions. It was discussed less extensively with the executive because the focus of our contract was more directly on Tony's leadership style than on aspects of the school programme for which the executive were responsible. In addition, the final version of the theory was developed so late in the process that we simply ran out of time. One of the factors that is seldom reported in research, but which has a huge bearing on the way it is conducted, is the resources available. Perhaps it is time to declare that the two case studies were done over a period of three years on a budget of $NZ 15,000, with no research assistance other than that provided by a transcription typist. The data collection, analysis and intervention processes were all personally conducted by the researchers, including the checking and analysis of sixty transcribed tapes.

The same criticism can be made of the problem analysis of the second case as was made of the first. While it explains the problems associated with Tony's style, it does not show how that style may, in part, be a response to the culture of the school which he was trying to change. His staff were deeply divided in their educational philosophy along a conservative–liberal continuum, and his impartiality was an attempt to deal with that division. With more time, resources and insight, we could have extended our theory-in-action analysis to embrace the key understandings and strategies that staff employed to deal with their principal and the changes he initiated, and evaluated that theory in terms of staff's ability to engage with their leaders in ways that enabled them to learn about the problems which they detected in the transition process. Whether or not these omissions turn out to be critical to the effectiveness of our theories will be judged in Chapters 7 and 8.

Problems in Data Collection and Analysis in PBM

The Scope of Data Collection

Some years ago I had a discussion with a colleague, who also taught qualitative research methodology, about students' difficulties in defining the boundaries of their data collection efforts. If their project involved an ethnography of the local high school, for example, should they get to know the community service clubs, hang out in the students' favourite coffee bar, and cruise the town on Saturday nights as part of their data collection activities? My initial response to this question was to say somewhat smugly that ethnographers had trouble establishing boundaries of relevance because their research was not problem-focused. This answer was unsatisfactory, however, because both PBM researchers and ethnographers address research problems of the ill-structured variety and, therefore, relevance is determined in both cases by successive hypotheses about the scope and structure of the problem. The difference between the problem analysis processes of PBM and ethnography lies in the nature and scope of the problems that are typical of each approach. The typical ethnographic problem involves portraying the culture of a particular group of socially interacting individuals by "discerning how ordinary people in particular settings make sense of the experience of their everyday lives" (Wolcott, 1988, p. 191). The typical problem of PBM, by contrast, is not to explain a culture, but to explain how that culture resolves problems in the neutral sense in ways that may produce problems in the negative sense. Given that only limited aspects of the culture may be involved in the particular problem under investigation, the scope of the PBM problem is usually much more limited than that of the ethnographer. The difference can be seen clearly by comparing what would be involved in doing an ethnography of the Western College professional development programme with what was involved in showing how that culture produced the problems of low impact and staff resistance. The former research problem requires a description of all its features and a much broader search for the culturally shared meanings associated with it. The latter problem sets much tighter constraints on data collection, because the phenomenon to be explained was itself much more limited. The initial statements of the problems at Western College suggested that data should be collected on staff attitudes and on the effects of the programme on teaching and administrative practices. Once these data were collected, hypotheses about why they patterned as they did provided further guidance about what was relevant. Hence data collection activities fanned out into the selection, design and evaluation of programme activities.

The tighter problem focus of PBM helped rule out, as well as rule in, some data collection activities. For example, observations of professional development activities in other schools were judged irrelevant because we had no reason to believe that they would contribute, beyond our current data, to an explanation of the problems of low impact and staff resistance. It is much more likely that these observations would have been included in an ethnography of the programme, because several staff, including the principal, had participated in professional development activities in other schools. They would have been judged necessary to a "holistic reconstruction of the phenomena investigated" (Le Compte & Goetz, 1982, p. 54). In fact, some ethnographers take such a holistic perspective to their explanatory task that they would reject the search for a distinction between relevant and irrelevant data. As Wolcott (1988, p. 204) puts it, everything is relevant, it is just a question of how one can make things manageable.

> Researchers have a tendency (and, realistically, an obligation) to oversimplify, to make things manageable, to reduce the complexity of the events they seek to explain. Ethnographers are not entirely free from this tendency; if they were they would not set out to reduce the accounts of human social behaviour to a certain number of printed pages or to a reel of film. But they remain constantly aware of complexity and content. There are no such

things as unwanted findings or irrelevant circumstances in ethnographic research.

While Wolcott recognises the need to make things manageable, his statement about relevance sheds no light on how this may be rationally accomplished. I suggest that in both PBM and ethnography it is judged according to successive hypotheses about the scope and structure of the problem, and that given the generally larger scale and looser constraint structure of the typical ethnographic problem, this will be a more uncertain and difficult task in ethnography than in PBM. Given the practical purposes of PBM, its relative manageability is a considerable virtue.

The problem focus of PBM also guides the way one checks for bias in data collection activities. In Northern College, for example, our sample of staff to be interviewed was drawn on the basis of their likely attitude, as predicted by two informants, to Tony's moves to restructure the school's management systems, and not on the basis of the usual biographical characteristics. The theoretically relevant dimension, from the perspective of avoiding sampling bias, was staff's educational philosophy, not their gender, age or years of experience. Similarly, in Western College, our data would have been biased if we had only talked to staff who were resistant to aspects of the programme, or if we had only attended those meetings in which Carol was likely to be more controlling. The question of bias in data collection is determined in PBM by whether or not one has collected data from sources that are likely to disconfirm as well as those that are likely to confirm current evaluations and hypotheses.

The above sampling strategy contrasts with that advocated by some ethnographers. Ball (1990, p. 165), for example, advises ethnographers to look for sampling bias by checking for an uneven distribution of data collection effort across times of day, places, people and activities. Similarly, Le Compte and Goetz (1982, p. 42) advise ethnographers to maintain an adequate inventory so that their findings are representative of all participants and circumstances. In PBM, any such unevenness may be entirely appropriate, given that events and activities relevant to the problem are disproportionately distributed across these dimensions.

The Problem of Data Reduction

Matthew Miles, a well-known practitioner of qualitative research, wrote with disarming candour some years ago about the difficulties associated with attempts to discipline the mountains of data that are likely to be accumulated in the course of a qualitative study (Miles, 1979). Although a respondent to the article attempted to point out that the weaknesses of qualitative research were not as insurmountable as Miles had portrayed (Yin, 1981), the latter's article drew attention to the ways in which the analysis process can become tedious and highly idiosyncratic.

Despite establishing boundaries of relevance for data collection, the quantity of data collected in the two cases was far greater than that reported. How did we discipline it in the process of developing our problem analyses? Why were some data given far more attention than others? Data reduction was achieved by searching through transcripts of meetings and interviews for instances of behaviour that represented classes of interest. Behaviour became interesting because by its membership in a class it could be accorded some theoretical significance—it could stand for something other than itself. This immediately raises the question about what makes a class of behaviour interesting. Are these classes brought to the investigation by the researcher or do they emerge during the analysis process? The answer is that there is an interplay between the theorising of outsiders and insiders in the development of these classes. At Western College, for example, transcripts of Carol's approach to problem-solving were searched for exemplifications of unilateral control and critical dialogue because we brought hypotheses to the research about the importance of these forms of interaction to problem-solving effectiveness. On the other hand, the category of behaviour and procedures called "meeting needs" emerged from the language of staff themselves and was adopted by the researchers as they realised its significance for the selection and design of professional development activities.

Whether or not a class of behaviour remains of interest depends on the contribution it makes to the explanation of the problem. Classes of interest must be woven together in ways that show how they produce the problem, so further data reduction occurs as initially interesting evidence falls outside the emerging theory of the problem. For example, in an earlier draft of the first case, the formality of the consultations between the executive and staff was put forward as a reason for the apparent lack of impact of the PDC sessions. Perhaps information would be more readily available, and motivation stronger, if consultation was seen as a matter of on-the-spot assistance rather than of a formal interview. Subsequent analysis of the PDC data suggested, however, that it was not their formality or informality that was contributing to their ineffectiveness, but the perceived threat associated with close examination of a colleague's practice. As a consequence of this reanalysis, the previously interesting class of formality–informality disappeared from our writing.

These procedures for the selection and hence reduction of data contrast with those based on the detection and description of regular patterns of behaviour (Van Maanen, 1979, p. 539). Under Van Maanen's procedures, relative frequency becomes the criterion for data selection. This principle is appropriate for those who wish to provide a comprehensive description of a phenomenon, but has several limitations in the context of a problem analysis. First, there were numerous regularities of belief and action associated with the professional development programme, which were not attended to because it was not clear how they could contribute to an understanding of the problem. For example, the lunchtime meetings of the professional development committee were punctuated by light-hearted banter about their work. The various professional development activities that the researchers attended were superbly organised, with everything from details of seating to group facilitation and introductions taken care of with efficiency and courtesy. Although several patterns can be detected in these data, they were not selected as classes of interest and incorporated into the analysis because they did not seem relevant to the explanation of the focus problems. Second, some behaviours which occur with very low frequency turn out to be important, because they hold special significance for those in the setting. For example, Tony's early departure from school on a couple of occasions to look after his young children was very important to some staff, because they saw it as symbolic of his lack of commitment to the school. In short, some behaviours which occurred frequently were not interesting, and others which occurred very infrequently were highly interesting. Hence our rejection of data selection principles based on relative frequency.

The Implications of Iterative Processing

The analysis of ill-structured problems involves repeated cycles of development, testing and revision of theoretical constructs and hypotheses. Given this iterative quality, PBM researchers cannot employ a methodology that requires constancy in definitions, categories and hypotheses. They must be able to remain open to rethinking the categories they employ and the interrelationships between them. At the same time, such openness brings with it costs in cognitive complexity, uncertainty and resources. The result is that the researcher walks a tightrope between premature closure of theoretical categories and their interrelationships and a costly search for new categories and relationships. Donald Schön (1983) has described how the practitioner and the researcher of practice need to exercise a kind of double vision in managing this process.

> At the same time that the inquirer tries to shape the situation to his frame, he must hold himself open to the situation's back talk. He

> must be willing to enter into new confusions and uncertainties. Hence, he must adopt a kind of double vision. He must act in accordance with the view he has adopted, but he must recognize that he can always break it open later, indeed, *must* break it open later in order to make sense of his transaction with the situation. This becomes more difficult to do as the process continues. His choices become more committing; his moves more nearly irreversible. As the risk of uncertainty increases, so does the temptation to treat his view as the reality. Nevertheless, if the inquirer maintains his double vision, even while deepening his commitment to a chosen frame, he increases his chances of arriving at a deeper and broader coherence of artifact and idea (p. 164).

As more is learned about the problem, both the way data are recorded and the way they are subsequently analysed changes, so that early efforts may be qualitatively different from later ones. Given that researchers only record what they notice at the time, and that what they notice changes with increased understanding, it follows that field notes taken at different stages of a project may be not directly comparable. The absence of evidence for a particular construct in early notes is as likely to be a function of changes in researchers' conceptual frameworks as it is of genuine shifts in the functioning and dynamics of the phenomena under investigation.

This iterative process has similar implications for data analysis. As more is learned about the problem, the meaning and definition of the categories will change so that analyses of early data sets will not be comparable to analyses performed on subsequent data. In theory, the researcher is left with the daunting task of having to review the whole data set each time the hypotheses change. The earlier quote from Donald Schön speaks eloquently about how this becomes increasingly costly as investment in prior formulations increases. The practical and financial constraints on repeated searches through data to develop and test new problem analyses are enormous. How then can the analyst be both open to new understandings and manage the process of testing a series of hypotheses about the nature of the problem? What is required is an efficient process for testing successive hypotheses until a theory emerges which explains the way numerous variables work together to form a problem. Take the previously discussed example of the hypothesis about the relationship between the formality and impact of professional development consultations. Informal testing of the hypothesis involved critical examination of its logic, and a search for disconfirming examples, by reviewing our data to see whether critical evaluation was more likely to be expressed in an informal than in a formal context. This process did not require us to review all our data; only to find a few disconfirming cases where people failed to make the distinction we were postulating, or failed to act as we had predicted in the more informal professional development activities.

Some qualitative researchers might object that these informal testing procedures are far too unsystematic; that in reviewing our data we could easily overlook the disconfirming examples, and that investment in a systematic coding process would have strengthened our analyses. Given that many qualitative researchers employ such coding schemes (Miles & Huberman, 1984; Strauss, 1987), and that we ourselves began to develop one at the beginning of our project, it is worth reviewing the merits of our decision not to continue with this task. Our reasons for abandoning it were that the level of precision required to develop a reliable coding system about leadership style, for example, seemed mismatched to the purpose of developing a theory of the problem. Every time we finished coding a transcript we realised we had altered our understanding of how the theory applied in the context of the case. In order to maintain comparability with previous transcripts, we should have recoded those that had already been completed. This task was far beyond our resources, but, more importantly, we realised that the employment of a formal coding system presupposes that which we were setting out to develop—namely a theory of the problem. We could only stabilise our codes, calculate levels of inter-observer agreement and quantify the occurrence of classes of interest once we were sure of the structure of the problem; that is, once we had made an ill-structured problem well-structured.

Perhaps the tension between iterative processing and efficient data management can be partly resolved by employing low-level descriptive categories which are kept constant throughout a study, and which facilitate the retrieval of data relevant to the development of more sophisticated theoretical distinctions (Le Compte & Goetz, 1982, p. 41). For example, data relevant to testing a hypothesis about the openness of Carol's leadership style could have been accessed via a "leadership style" code, which did not prejudge what qualities of her reasoning and action counted as "open" and "closed". If such a code encompassed more data than could be easily reviewed, it could have been broken down into subcodes that reflected different data contexts, and thus the requirement for efficient data management could still have been met without sacrificing theoretical flexibility.

Generalisation in PBM

A final concern about the analyses presented in Chapters 4 and 5, and about PBM in general, might be about the generalisability of its knowledge claims. While the primary purpose of PBM is the understanding and resolution of problems of practice, it is not clear whether any such understanding would apply beyond the particular problem under investigation. One could argue that PBM does not yield generalisable knowledge, because it is typically concerned with the study of particular problems or single cases rather than with population samples. Evaluation of this argument requires an understanding of how it is that the logic of sampling theory enables generalisations to be made, and a decision about whether sampling is a necessary condition of generalisation.

The argument from sampling theory can be made through a hypothetical research example. A researcher wishes to test a hypothesis about the effects of a democratic leadership style on problem-solving effectiveness. An experimental problem-solving situation is designed, in which psychology undergraduates are randomly assigned to roles as members and leaders of democratic and autocratic groups. The results show that groups with democratic leaders score more highly on problem-solving than groups with more autocratic leaders.

The logic of sampling theory suggests that the demonstrated causal relationship between leadership style and problem-solving effectiveness can be generalised to the population of subjects, settings, treatments and outcomes which are represented in the study. We can examine this logic in terms of the single dimension of generalisation across subjects. The students used in the experiment have been carefully described on variables of age, intelligence, social class and years of education. Generalisation to subjects of similar characteristics is warranted, not because these variables have a causal connection to problem-solving effectiveness, but because they are thought to be strongly correlated with those, as yet undescribed variables, which play a direct causal role. It is this association between alleged causal mechanisms and variables acting as surrogates for a complete explanation of those mechanisms which underpins the logic of this type of generalisation.

If sampling designs provide the *only* basis for generalisation, then PBM cannot produce generalisable findings. There is, however, a rival conceptualisation of generalisation which is far more promising for PBM because it does not rely on sampling procedures. Cook (1991) contrasts the view that extrapolation of research findings is achieved through sampling (a view he associates with Donald Campbell) with the view that it is achieved through causal explanation (a view he associates with Cronbach). He describes some of the issues this way:

> Cronbach assumes that extrapolation is best achieved through causal explanation rather than sampling theory. He wants evaluators to learn *why* a treatment is or is not effective rather than identifying *whether* it is effective. He seeks explanation through several different mechanisms, including: decomposing macro-level variables to identify their causally efficacious components; learning which molar forces statistically interact with the treatment; and identifying those forces that temporally mediate between when a treatment is manipulated and when an effect is observed (p. 130).

In short, Cronbach sees a sampling approach to generalisation as counterproductive because it "focuses attention on superficial correspondences rather than generative causal processes" (Cook, 1991, p. 130).

In understanding how causal theorising can be an alternative route to generalisation, it is probably useful to think about those branches of science which are widely seen to be highly productive of generalisable knowledge claims, yet which seldom, if ever, employ a sampling methodology. Neurophysiologists typically claim causal links between certain forms of brain stimulation and perceptual response without having sampled the animals or humans on which they have completed their research. They claim species-wide applicability for their neural and perceptual findings, not because they have sampled the population of laboratory cats, but because their particular findings cohere with a wider set of relevant theories which are accepted as valid descriptions of species' functioning. Until these theories are substantially altered or refuted, the new findings will enjoy whatever generalisability is granted to the theories in which they are embedded.

If causal theorising is an alternative to sampling theory in the quest for generalisable knowledge, then the next question to ask is about the potential of PBM to yield such causal theories. Once again the traditional answers are fairly pessimistic, because PBM lacks those features of inquiry that are thought necessary to establish causal relationships. Those features are associated with the traditional Fisherian design, in which the combination of pre- and post-treatment observations and random allocation to treatment and no treatment groups increases the confidence with which we attribute causal influence to introduced treatments (Cook, 1991, pp. 115–116).

As we shall see in Chapter 7, the interventions associated with PBM are not introduced under experimental or even quasi-experimental conditions. PBM's contribution to causal theorising depends on finding other ways to achieve what experimental and quasi-experimental designs do relatively successfully; namely, rule out competing explanations of observed difference. Once again, however, if one examines the logic underlying experimental techniques, one can be more sanguine about the potential contribution of PBM to causal theorising. Suggested causal accounts of practical problems and of observed changes gain validity to the extent that the researcher raises and replies to competing explanations. As I have already discussed, intensive involvement in the practice context, and the competing theories of practitioners, are a rich source of plausible alternative hypotheses; in addition, the intricate patterning of the qualitative data typical of complex problems sets tight empirical constraints on causal theorising (Campbell, 1978). Granted, such hypothesis testing strategies leave one uncertain of the plausibility of the resulting theories, but neither do the more rigorous experimental and quasi-experimental designs rule out all threats to the validity of causal accounts (Cook, 1991).

My conclusion, therefore, is that researchers employing PBM or other methodologies which do not typically involve sampling will contribute generalisable knowledge to the extent that they move beyond the local to seek the potentially universal, by linking their analyses (through challenge or confirmation) to already accepted generalised knowledge claims (Robinson, in press). This argument suggests that the generalisability of the theories developed in Chapters 4 and 5 depends on both the prevalence of the problems they address and the extent to which the researchers' explanations draw on mechanisms which are known to operate in similar situations.

With respect to the first condition, the problems of the Western College professional development programme occur in many interpersonal and organisational contexts where there is a requirement to resolve tough problems and to do so without destroying collegiality and teamwork. There is a descriptive literature which documents this type of problem in the area of teacher development and accountability (Bridges, 1986), negative feedback (Robinson & Timperley, 1989), employee appraisal (Beer, 1987) and consultation (Rossmore, 1989). There is less evidence about the prevalence of the type of problem investigated at Northern Grammar, because most of the literature is concerned with the theory of democratic management (Pateman, 1970; Rivzi, 1989) rather than with the difficulties of achieving it.

With regard to the second condition for generalisation, the problems were explained with reference to aspects of interpersonal and organisational functioning which have been widely investigated and shown to have consequences similar to those established by other studies. Argyris, for example, has written extensively about the consequences of

Models One and Two for interpersonal and organisational learning (Argyris, 1984, 1985) and both he and Rossmore (1989) have written about the dynamics of threat avoidance. In short, although the theories were tailored to the specifics of the focus problems, and their credibility depended in part on the way they captured local conditions, their links to more abstract and well-documented features of interpersonal organisational theories of action greatly enhance their generalisability. Any leader who attempts to implement a passionately held vision in a politically divided organisation by staying impartial and low key during the implementation processes will, if our theory is correct, experience the consequences which Tony did at Northern Grammar. Any professional development programme which is premised on the assumption that effectiveness requires threat avoidance will be impotent in the face of tough issues. These cases speak to other similar situations because the theories of action that explain them are not idiosyncratic. Since they exhibit the generic features of those theories of action that we already know inhibit learning and problem-solving, our knowledge of those particular cases can be linked to our knowledge of all cases in which this type of theory of action operates.

Summary

The first half of the chapter described the steps taken to improve the explanatory accuracy of the problem analyses presented in Chapters 4 and 5. These steps included deliberate searches for alternative hypotheses, and for disconfirmation of currently favoured ones, along with the involvement of research participants in a process of member checking. This latter procedure was described in detail to show how the PBM approach differs from that advocated by Lincoln and Guba (1985) in its far greater emphasis on the achievement of agreement about the warrant for inferences and conclusions, rather than the achievement of agreement *per se*. In addition, it was argued that given the practical problem-solving purpose of PBM, such agreement must be achieved at all levels of analysis, including that of the researchers' theoretically informed critique of the participants' understandings and actions.

The second half of the chapter included reflections on selected aspects of the process of problem analysis. The first involved the way decisions are made about what data to collect. A comparison of the decision process in PBM and ethnography showed that while initial hypotheses about the scope and structure of the problem guide decisions about relevance in both approaches, the smaller scope and different nature of the problems addressed in PBM provide tighter boundaries of relevance than are typical in ethnography.

The processes of data selection and reduction were described in terms of a search for behaviours in classes of interest, and the weaving together of such classes to show how they combine to produce a problem. This principle of data reduction was compared with that based on the detection of regular patterns of meaning and behaviour; a principle that is rejected in PBM because many regularities are unrelated to the focus problem, and some low-frequency evidence turns out to be highly significant. The implications of the iterative processing that is a feature of ill-structured problem-solving for the development of systematic, yet flexible approaches to data analysis were also discussed.

Finally, the approach to generalisation based on the study of population samples was contrasted with that based on causal reasoning. It is this latter approach which makes it possible for PBM to yield generalised knowledge, despite its focus on particular problem situations. Generalisable knowledge is produced, to the extent that the analysis of particular problems links with more abstract theoretical knowledge and principles, known to be applicable across a wide range of similar situations.

References

Argyris, C. (1983). *Reasoning, learning and action*. San Francisco: Jossey-Bass.

Argyris, C. (1985). *Strategy, change and defensive routines*. Marshfield, MA: Pitman Publishing.

Ball, S. J. (1990). Self-doubt and soft-data: Social and technical trajectories in ethnographic field work. *International Journal of Qualitative Studies in Education*, **3** (2), 157–171.

Beer, M. (1987). Performance appraisal. In J. Lorsch (Ed.), *Handbook of organisational behavior*, pp. 286–300. New York: Prentice-Hall.

Bridges, E. M. (1986). *The incompetent teacher* (Stanford Series on Education and Public Policy). Lewes: Falmer Press.

Campbell, D. T. (1978). Qualitative knowing in action research. In M. Brenner, P. Marsh & M. Brenner (Eds.), *The social contexts of method*. London: Croom Helm.

Cook, T. D. (1991). Clarifying the warrant for generalized causal inferences in quasi-experimentation. In M. W. McLaughlin & D. C. Phillips (Eds.), *Evaluation and education: A quarter century*, pp. 115–144 (Ninetieth Yearbook of the National Society for the Study of Education, Part II). Chicago: University of Chicago Press.

Cronbach, L. J. (1980). Validity on parole: How can we go straight? *New Directions for Testing and Measurement*, **5**, 99–108.

Le Compte, M. D. & Goetz, J. P. (1982). Problems of reliability and validity in ethnographic research. *Review of Educational Research*, **52** (1), 31–60.

Lincoln, Y. & Guba, E. G. (1985). *Naturalistic inquiry*. Beverly Hills: Sage.

Miles, M. B. (1979). Qualitative data as an attractive nuisance: The problem of analysis. *Administrative Science Quarterly*, **24**, 590–601.

Miles, M. B. & Huberman, A. M. (1984). *Qualitative data analysis: A sourcebook of new methods*: Beverly Hills: Sage.

Pateman, C. (1970). *Participation and democratic theory*. Cambridge: Cambridge University Press.

Phillips, D. C. (1987). Validity in qualitative research: Why the worry about warrant will not wane. *Education and Urban Society*, **20** (1), 9–24.
Rivzi, F. (1989). In defence of organizational democracy. In J. Smyth (Ed.), *Critical perspectives on educational leadership*, pp. 205–235. Lewes: Falmer Press.
Robinson, V. M. J. (in press). Current controversies in action research. *Public Administration Quarterly*.
Robinson, V. M. J. & Timperley, H. (1989). Research and practice in negative feedback processes: The limitations of a model of unilateral control. Unpublished manuscript.
Rossmore, D. (1989). Leader/consultant dilemmas: The primary barrier to satisficing. *Consultation*, **8** (1), 3–24.
Schön, D. (1983). *The reflective practitioner*. New York: Basic Books.
Strauss, A. L. (1987). *Qualitative analysis for social scientists*. Cambridge: Cambridge University Press.
Van Maanen, J. (1979). The fact of fiction in organizational ethnography. *Administrative Science Quarterly*, **24**, 539–550.
Van Maanen, J. (1982). Fieldwork on the beat. In J. Van Maanen, J. M. Dabbs & R. R. Faulkner (Eds.), *Varieties of qualitative research*, pp. 103–151. Beverly Hills: Sage.
Van Maanen, J. (1988). *Tales of the field: On writing ethnography*. Chicago: University of Chicago Press.
Van Maanen, J., Dabbs, J. M. & Faulkner, R. R. (1982). *Varieties of qualitative research*. Beverly Hills: Sage.
Wolcott, H. F. (1988). Ethnographic research in education. In R. M. Jaeger (Ed.), *Complementary methods for research in education*, pp. 187–206. Washington, D.C.: American Educational Research Association.
Yin, R. K. (1981). The case study crisis: Some answers. *Administrative Science Quarterly*, **26**, 58–65.

7

Intervening in Educational Problems: Two Case Studies

People change when they believe that their current theories about the world are unsatisfactory in some way and when they are confident of an alternative. This suggests that the problem-resolution purposes of PBM are not achieved through critique alone; that the PBM researcher must also be an interventionist who can provide, or work with practitioners to develop, a more adequate theory of action. This chapter explores some of the implications of this interventionist role by describing and evaluating the intervention phase of the two cases. The intervention theme is continued in the subsequent chapter by critically reflecting on the cases, and on the intervention potential of PBM itself. The broad question I am pursuing is, "How does one help people to change their theories of action when those theories are implicated in educational problems?"

Our problem analyses suggested that the theories of action employed in the two cases fell short on the criteria of effectiveness, coherence and improvability. The effectiveness evaluation suggested that relevant goals would not be achieved; the coherence evaluation suggested that even if they were, other problems would be created or exacerbated; the improvability evaluation suggested that the principals would be unable to solve these problems on their own. We had to decide whether we would just address the ineffectiveness, or whether we would also address improvability through the much more challenging and difficult issue of the quality of the principals' own problem-solving processes. At Western College, choosing the former would involve advising Carol and her executive staff on a more effective design for a professional development programme and helping them to implement it. Choosing the latter would involve helping them to understand why they had designed an ineffective programme, why they had had difficulty improving it, and how they could design one that was more effective. Similarly, choosing the first option at Northern Grammar would involve providing Tony with more effective solutions to his transition problems; choosing the second would involve helping him to understand why his current problem-solving style was incompatible with achieving his vision for the school, and helping him to learn a more effective approach.

There are several general grounds for choosing the second option. First, focusing on practitioners' problem-solving processes is justified when they might be reasonably expected to solve the problem that concerns them. In the case of professional development, for example, principals of New Zealand schools are now expected to develop programmes of staff development and appraisal, and some attempt is made by the external review agency to see that this is done (Department of Education, 1988). In other words, there is now an expectation that New Zealand school leaders learn how to manage appraisal and professional development programmes, including any problems that arise within them. Second, a focus on problem-solving processes is also appropriate when the nature of the problem and hence of the solution is unstable, and therefore requires constant monitoring, evaluation and adjustment of solution strategies. Given such instability, practitioners who lack adequate inquiry skills will be unable to make the adjustments required to sustain the long-term effectiveness of a solution strategy. Finally, a focus on inquiry values and skills is appropriate when those values are essential to effective implementation of a new solution. When solutions conflict with existing theories-in-use, inquiry values ensure that the points of conflict are resolved through testing the relative merits of the two theories, rather than by the relative power of those who advocate for and those who oppose the innovation.

These reasons were behind our choice to focus on problem-solving processes at Western College. First, the principal and her senior staff acknowledged their increased responsibility for the professional development and appraisal of their own teachers. Second, we knew that Carol's lack of inquiry skills had made it difficult for her to understand the resistance of those staff who held different attitudes to appraisal and professional development. If Carol did not change the way she handled this resistance, our intervention could make it worse, because the type of programme we were advocating would bring these differences increasingly into the open. If we were not to collude with Carol in imposing a solution on staff, we needed to teach her how to involve them in testing her beliefs about its desirability, just as we had involved her in testing ours.

At Northern Grammar, similar arguments applied. We had to address Tony's problem-solving skills, because it was these that were jeopardising his attainment of his vision of democratic management. The ineffectiveness of his theory of transition (Figure 5.1) was due to the dilemmas that he experienced about how to inquire into, challenge and test emotionally-laden staff disagreement. Until we altered the values and understandings that created these dilemmas, any advice that presupposed his ability to do this would be ineffectual.

The Feedback Process

We could not proceed to an intervention phase if Tony and Carol did not agree with our analyses. Our contract had been to investigate some problems, not to resolve them. Carol and Tony were unlikely to be moved to act differently if they believed the analysis was distorted by selective data collection or reporting, faulty inferences or incomplete understanding of the conditions under which they worked. Six months after we first started collecting data about the Western College professional development programme, we presented an interim analysis to the principal and asked for her reactions. Our objectives were to present our analysis to Carol, and subsequently to her executive staff, in a way which encouraged challenge and questions.

> *VR*: What we have done here Carol is to try to summarise where we are at, [. . .] in our thinking about professional development in the college. We fully expect this to be the first step in the process whereby you might want to take this material away, and reflect about it. You might want to confirm or disconfirm pieces of it as we go through it now. And we see coming to some approximate agreement about its meaning as a pre-requisite really, to then deciding what if anything might be a useful next step to take [WCTR30, 7.3.88].

At the end of the first feedback session with the principal, we had a working agreement about the broad accuracy of the analysis and about the next steps. Carol wanted to focus on her own skills in understanding and reacting to staff who were resistant to the programme, and we agreed with that step, because it was consistent with our wish to focus on the way Carol herself tackled the problems associated with the programme. The next three sessions involved a series of both practical and theoretical examinations of Carol's and the researchers' theories of problem-solving. The validity of the analysis of the researchers and the usefulness of their bilateral alternative were constantly tested in a series of role-played and real conversations with staff who had resisted involvement in the programme. If our analysis was correct, then Carol would not be able to talk with these staff in ways that left them and her feeling that progress had been made. If our theory was a viable alternative, then we should be able to do so in ways which both parties recognised as different and more productive.

Carol found these practical demonstrations compelling enough to want to learn about the theory that informed our practice. We can see from the following extract, however, that at this early stage she understood us as offering her a more effective set of techniques for winning staff over, rather than as challenging her to reframe "resistance" as an opportunity to enter into critical dialogue with those who did not share her assumptions about professional development. After one such demonstration with a "resistant" staff member (Chris), Carol explained her reactions as follows:

> *VR*: Is there anything you want to say about what's happened so far?
>
> *Carol*: I thought I'd find it [the researchers' demonstration] disturbing, actually, but it's not. Very nice, laid back, Yeah.
>
> *VR*: Oh, that's nice.
>
> *Carol*: I wasn't sure what my reaction would be.
>
> *MA*: Has it felt useful?
>
> *Carol*: Mmmmmmm. Believe me that's a year's work you've done for me. Literally. Got further in your seven minutes than we did all of last year.
>
> *MA*: That's with Chris. What about for you? What happened? You did your own interview with Chris, and saw Viviane's role-play with Chris. What was in there for you? As opposed to getting something out of Chris? [WCTR32, 22.3.88].

We presented our feedback to Tony in the form of a simulated conversation between two staff members rather than in a written form, as with Carol, because we believed that this would be a more effective way of communicating some of the emotional issues that had emerged from our interviews with staff. The feedback was a collage of staff reactions to Tony and to the changes he had introduced, and was designed to communicate staff's sense of puzzlement, wariness and uncertainty, as well as some specific positive and negative reactions. Tony's immediate reaction was that little of the feedback was a surprise, but that he would need time to listen to the tape and digest it fully. We were to discover later that he had suppressed his emotional reactions in order to cope, and that he was both angry and in disagreement with much of what we had presented. The way we resolved this disagreement and its implications for intervention in PBM is discussed at length in the next chapter. The eventual result, however, was a new contract in which he agreed to join Carol in a series of workshops designed to teach the theory and practice of critical dialogue and to apply those skills to the achievement of his goals for the school.

The Intervention Process

Our intervention sequence began with teaching Tony and Carol the theory and practice of critical dialogue. Later, with the encouragement of the principals and their executives, we included the latter in training workshops in which the principals worked with us as co-leaders to train their executives and to address school-related problems. We justified starting with the principal by arguing that more junior staff were unlikely to take the risks involved in critical dialogue, unless their boss had first demonstrated a sincere commitment to its underlying values. In addition, the final responsibility for effective school management lay with the principals not the researchers, so we were keen to support them in their problem-solving efforts, rather than to take over that role ourselves (Block, 1981).

The particular intervention processes employed in the two cases are based on those of Chris Argyris and his colleagues (Argyris, 1970, 1976, 1985; Argyris & Schön, 1978). They were chosen because they address the way people reason and act in a problem situation, and because my colleague and I possessed a modest level of expertise in their use. I do not wish to imply that these are the only intervention processes suited to the action phase of PBM. Any intervention process which reconstructs, evaluates and improves theories of action is potentially consistent with this methodology.

The Content of the Workshops

The researchers brought the theory and practice of critical dialogue to the workshops; the principals brought their competing theories and the objections that those generated. We began by teaching the three basic rules of critical dialogue, and showing how they contributed to interpersonal and organisational learning and problem-solving (Table 7.1). These basic rules involve disclosure of your opinion and reasoning processes and checking whether others perceive things as you do. They serve the values of increasing the validity of information and the choices available, because they encourage checking and the expression of difference.

Readers might be tempted to ask why an elaborate learning process is needed to teach people some skills which are already in their repertoire. After all, all normal adults know how to express their opinion, how to back up that opinion with evidence or argument, and how to ask for the reactions of others. The answer to this question lies in the fact that the use of these skills is frequently incompatible with the

TABLE 7.1. *Three Rules of Critical Dialogue*

	Rules	*Examples*
1.	Say what you think.	"I want you to take some responsibility for helping me get out of this mess . . .
2.	Say why you think it.	. . . because I thought you had agreed to help me meet this deadline."
3.	Check with others.	Do you remember it that way?

TABLE 7.2. *Unilateral Control as a Barrier to Critical Dialogue*

	Critical dialogue	*Unilateral barriers*
1.	Say what you think.	Hide what you think when you think it might upset others. Hide what you think when you privately judge that saying it is not to your own advantage.
2.	Say why you think it.	Don't say why you think it because the correctness of your view is obvious. Don't give reasons when you expect compliance—"because I say so". Don't give reasons when you're not confident of their worth.
3.	Check with others.	Don't check when persuading others. Expect others to think as you do since your views are obvious and right.

unilateral values of winning, and doing so without hurting others. Table 7.2 lists some of the ways in which those values act as barriers to the use of critical dialogue.

When people are thinking in unilateral ways they will not use the skills of critical dialogue, even though they have them in their repertoire. Learning to engage in critical dialogue takes time and is frequently painful because it requires the unlearning of the habitual patterns of reasoning illustrated in the right-hand column of Table 7.2. If critical dialogue is reduced to a package of communication techniques separate from its value base, it becomes a tool for the manipulation of others, because unilateral reasoning processes which are designed to protect oneself from influence remain intact.

The importance of unlearning unilateral styles of interaction in order to learn critical dialogue was a constant theme of our workshops. The first chance that Carol and Tony had to assess the extent to which they

TABLE 7.3. *Time Spent Teaching Critical Dialogue (hours)*

	Western College	*Northern Grammar*
Individual sessions	11	7
Joint training of principals	11.5	11.5
Executive staff training	12	15.5
Total for each school	34.5	34.0

employed unilateral strategies came with their completion of the standardised Ted–Don negative feedback scenario, which was discussed along with Carol's response in Chapter 2. Tony's response was equally unilateral as that of Carol, though the techniques he used to persuade Ted of the validity of his feedback tended to be more indirect. Several sessions were spent discussing this exercise, so that Tony and Carol could check the way the researchers had analysed their responses and begin to understand the theories of unilateral control and critical dialogue in the context of their own practice.

THE PROCESS OF THE WORKSHOPS

The theory and practice of critical dialogue were taught through individual discussion with each principal, and through a series of workshops for both principals and the executive of each school (Table 7.3).

Several principles guided our design of each session. First, we wanted to use real school issues as the material of the course, so that difficult questions of transfer to the real world were dealt with in the training itself. At several of the individual sessions, for example, we involved staff members whom the principals believed held very different views from their own on a relevant issue. The focus of these sessions was the principals' ability to listen to the views of their staff, and to understand and respond to their concerns in a way that left both parties feeling more satisfied about their relationship, and about their ability to tackle the problem, than they had previously. This was achieved by discussing, modelling and coaching the principals in the reasoning and interpersonal skills involved.

A second important principle for the workshops was to spend as much time as possible producing and examining what people actually said and did, rather than what they thought they said and did, so that we could monitor the congruence between theory-in-use and espoused theory. This was achieved through various forms of experiential learning such as simulated and real conversations, video-taping and the analysis and discussion of transcripts of these activities.

A third principle was the importance of participants' objections to critical dialogue. If the principals were to become intellectually committed to this alternative theory, then we had to help them surface and develop answers to any objections. The transcripts show that despite their struggle to understand the theory, both principals and senior staff raised numerous objections to it.

> Did critical dialogue conversations take too long?
> Did critical dialogue presuppose a shared value base?
> What did critical dialogue say about feelings?
> Was it manipulative?

Understanding and answering these objections took a long time, particularly when there was more than one staff member involved in the session. The objections were handled by identifying the assumptions on which they were based and examining their logic and accuracy. In the following extract the researcher challenges Carol's view that her more unilateral approach saves time, by referring to the evidence of her previous conversation with a "resistant" staff member (PL). The researcher's challenge leads Carol to reflect on the validity of her theory about "saving time".

> *Carol*: [My strategy] saves time.
>
> *VR*: Well, it saves time, yes. I was reflecting on whether in fact it does, because in this case you have had a goal, which was to convince him of something and maybe change his attitude, and I think you acknowledged that you didn't achieve your goal, so I think I'd want to question—
>
> *Carol*: Mmmm.
>
> *VR*: — your assumption that the tactical strategy does in fact save time. That in this scenario, at least, you didn't achieve your goal, and maybe it didn't save time either. [. . .]
>
> *Carol*: True. The assumption I've been using there is that if you tick off all the facts in the background, you've got them out of the way. That's not necessarily true. They can intrude later on and be just as time consuming as trying to get them out of the way at the beginning of the conversation.
>
> *PL*: Plus, we may in fact be no closer, I feel, to achieving the objective that you set [WCTR32, 22.3.88].

Table 7.3 shows that learning critical dialogue in the context of real school issues involves a considerable investment of time. While I believe that this time can be shortened with more expert intervention, the learning process cannot be reduced, for example, to a one-day in-service course. For most people, learning critical dialogue involves a fundamental rethinking and reshaping of their assumptions about what it is to be effective and how to get there. A sporting analogy is useful in conveying what is involved. The learning process is not the equivalent of the novice learning to play tennis. It is more like a veteran player of fifteen years learning to change the way he grips the racket, so that he can eliminate the backhand slices and cramped forehand shots that have for so long limited his success at and enjoyment of the game.

The Effectiveness of the Training

The first question about effectiveness is whether or not we made a difference to the principals' approach to solving difficult problems. If a positive answer is justified, we can then examine the effects of these changes in each principal on the school-based problems that were the focus of each case.

Impact on Carol

Carol's style was quite typical of the forthright version of unilateral leadership, in that she often said what she believed in ways that made it difficult for others to disagree. Our previous analysis of her style of chairing an executive meeting showed that she made strong uncompromising statements and tended to define unilaterally the nature of problems, solutions and problem consequences:

> "staff need to be told . . .".
> "people have just got to realise . . .".
> "they are just not communicating . . .".
> "There is no doubt about that . . .".

Alongside these unilateral patterns, we identified patterns that were consistent with critical dialogue and which Carol wanted to build on during the workshops. She approached the challenge of this new learning by setting her own goals and taking it a step at a time.

> *Carol*: I can see the wood for the trees now because that's focusing on just one thing, whereas before there were so many things to focus on that I was really getting lost and you sort of didn't know which to start with, and that's why I focused on checking. It felt to

me that that was something immediately I could do more of all the way through.

MA: So it's a matter of building on that now is it?

Carol: And now it's the avoidance of the unilateral solution giving and decision-making that's very hard to shed. Very.

MA: Well maybe that's the next thing that we can spend a bit of time looking at? [WCTR35, 17.6.88].

One way of assessing the impact of the learning seminars on Carol's leadership style is to compare the patterns identified in Chapter 4 with her style at subsequent meetings. We chose to analyse the meeting at which she negotiated revisions to the draft teacher appraisal policy, because this meeting would test Carol's ability to remain open on issues that were dear to her heart and which she had previously sought to resolve with inappropriate speed. The purpose of the meeting was to consult with Heads of Department about suggestions for changes to the policy which had been received from staff. Carol began the discussion by reading out a summary of the feedback received so far and then suggested a next step:

Carol: The question is, where do we go from here? After asking the question, what I've tried to do is to present a possible answer [WCTR53, 31.7.89].

Carol has clearly identified her answer as a possibility, not as *the* answer. She then identifies her hope that the involvement of numerous staff in the development of the new policy will change staff's perception that the programme belongs to her rather than to the staff as a whole:

Carol: I would like to believe that staff assumptions about PDC and appraisal structures being my ideas and initiatives alone have changed to some extent. That really is the assumption I'd like to check, so please help by letting us have some feedback and I'm going to welcome that in a very open and honest way. If people are still feeling that if they approach me with something difficult, I'm being difficult about accepting or hearing what you're trying to say I need to know that [WCTR53, 31.7.89].

Carol seeks feedback here on two levels. She wants to know whether staff see the programme as "hers", but at the same time acknowledges that past patterns of interaction may make it difficult for them to give her honest feedback. As frequently happens in such meetings, some staff turned to discussion of the topic without answering Carol's questions. She repeated her request and staff eventually gave her the feedback she sought.

*HOD*1: From my point of view, it doesn't matter to me very much where the statement comes from initially, I mean, if you've written it that's fine, that doesn't concern me at all, what only concerns me is two things: first of all what's in it, and secondly what sort of influence we can have upon it if necessary. I think those are the key questions for me. So the fact that you may or may not have written it, that's not an issue for me personally.

Carol: Can I then check, picking up from what John has said, how do people feel about the influence that they may have on the processes?

*HOD*2: I think the process is such that I can contribute to it by just coming and discussing it with you, and whatever views I was concerned about I could explain to you knowing that even they would be answered or they would be taken further.

Carol: Right.

*HOD*2: I feel that I've been consulted [WCTR53, 31.7.89].

Satisfied that staff would tell her their disagreements, Carol then made a final check of their willingness to proceed with the proposed scheme before turning to the specific recommendations. The transcript shows numerous examples of checking, of advocating her opinion in ways that encourage a reply, and a willingness to challenge staff who are raising issues that are beyond the agreed agenda.

While Carol's communication was considerably more bilateral in this meeting, leaders who are skilled in critical dialogue also advocate for school policies and procedures which promote values of openness, learning and mutual accountability. There were times in the meeting when staff advocated for procedures which could have protected them from such opportunities. For example, some wanted the section in the appraisal form on weaknesses or areas for improvement to be replaced by a section on targets for the coming year.

Exec. member: ... my suggestion was that anything that's negative, ... should be addressed ... if we're going to write it, in terms of targeting. In other words there are targets to improve, to do better in this or improve this, rather than saying "I am weak in this." It's a type of professionalness ... in that way, targets are positive [WCTR53, 31.7.89].

A Head of Department made a similar statement about the need to avoid discussion of weaknesses.

HOD: So, I don't like the idea of commenting on what can be done better or focusing on disclosing weaknesses. I would much prefer

> they came into setting my targets for the next year [WCTR53, 31.7.89].

Carol checked that the group wanted such an alteration and did not inquire into the possibility that such a change could reflect avoidance of discussion about perceived weaknesses in teaching or administrative work.

> *Carol*: So, am I right in perhaps writing that up by saying there is some unease about incorporating that. . . . So, the first suggestion is that it should be termed as setting targets rather than commenting on what can be done better or disclosing weaknesses. Is that approved? [WCTR53, 31.7.89].

While Carol herself has learned that a professional development programme that avoids tough issues will have limited impact, this lesson also needs to be debated with those who wish to protect themselves and others from these issues. If Carol does not challenge the possibility that some staff's desire for protection can undermine their own and others' learning, her bilateral communication skills will produce a programme to which all are committed but which is still impotent in the face of others' discomfort.

Impact on Tony

After giving Tony feedback about our analysis, we agreed to test its accuracy further by helping him to obtain feedback from his own staff, and by providing direct help with the resolution of any discrepant perceptions of this style. We therefore invited to the first individual training session an executive member who had indicated some concerns about the way Tony had handled the recent allocation of staff positions of responsibility. Tony had wanted to use the positions in a way that would establish a flatter school structure, and had not communicated his plans or his reasons until the various steps were already in place. The staff member suggested that such behaviour would work against the development of a shared vision of the school and lead to suspicion of his wish to be democratic. The example was an independent illustration of our earlier diagnosis of Tony's practice as sometimes manipulative in its attempt to reach democratic ends.

Tony was able to reflect on the factors that had prevented him from being open about his plans. They included his wish to avoid being lobbied, but more importantly his inability to deal directly with those he thought might be hurt by them. He kept his plans close to his chest because he did not know how to deal with the consequences of openness. This insight led to discussion about the focus of the next learning seminar.

> *VR*: That may be the sort of conversation that's very important for us to practise.
>
> *Tony*: Mmm. Yes it is. To be tough without hurting. Yes. It is something I don't own up to.
>
> . . .
>
> *Tony*: I accept being complete enemies of [some people]. I can take them off at the knees.
>
> *VR*: Yeah.
>
> *Tony*: But people I've got to work with. I don't know how to do it.
>
> *VR*: And I'm thinking that there's a connection between that difficulty you say you're having and your reluctance to share some of the long range issues. And if we're talking about resolving the dilemma in a way more like what [executive member] is suggesting—
>
> *Tony*: Yeah.
>
> *VR*: — that training's essential.
>
> *TS*: That would be very, very useful to me. Although I'm frightened about what I would then do with it.
>
> *MA*: Yeah.
>
> *Tony*: Yeah. I've talked to you about my fear. You know the thing that I don't ever want to do, I think I've told you about it. I don't want to become an autocrat.
>
> *VR*: Right.
>
> *Tony*: If I'm given the skills what will I do? I don't want it to be easy. At the moment I actually hate hurting people [NGTR31, 11.3.88].

At first we used situations which were not school related so that Tony could concentrate on the skills being taught, without becoming embroiled in the content. The next step was to apply them to real issues, and this required us to involve the whole of the executive staff.

By this stage there were several strands to the intervention phase of the research and it was not until two months later that we returned to Tony's skills in being open about tough issues. Our objective was to

TABLE 7.4. *Tackling Tough Issues*

Rules of critical dialogue	*Examples*
1. State what you are concerned or embarrassed about.	"I'm concerned about the amount of noise coming from your classroom . . ."
2. State the reason for your concern.	". . . because I don't think the students can concentrate in that atmosphere and I also worry about the effect it's having on you".
3. Help the other person to give their views.	What's your reaction to the noise?

teach a version of the basic rules of critical dialogue, adapted to fit the situation where difficult issues were being addressed (Table 7.4).

Transcripts of our training sessions show that while we were successful in teaching these skills, we were less successful in altering the reasoning processes that guided when Tony chose to use them. There were many occasions when he still withheld his concerns because he believed that the moment was not right, or because he believed his information base was not strong enough. In the following example, Viviane has told Tony that several staff are puzzled by his failure to challenge a senior staff member who disciplines students in a way that is inconsistent with his espousal of a caring school. Tony explains his actions by saying that the staff person in question:

> . . . almost never did it when I was around—like never after the first time [I confronted the staff member] but I know it would happen because I would hear about it weeks later or days later

TABLE 7.5. *The Reasoning Behind a Failure to Challenge*

Situation	*Actions*	*Reasoning*	*Consequences*
I say I care about students.		I justify my inaction by appealing to parts of the theory I have learned (need for good data).	Some staff see me as not caring about students despite my rhetoric.
AND	I choose not to challenge the staff person.		
I have information that a staff member is acting in uncaring ways.		I ignore those parts of the theory which would help me to obtain such data (disclosing, checking, and involving others).	Skills of critical dialogue are used in ways that maintain my existing reasoning processes.

> from somebody's mother or whatever. Always of course when it's impossible to reconstruct an event or do anything about it [NGTR61, 20.3.90].

Table 7.5 critiques this reasoning in terms of its compatibility with critical dialogue.

In summary, Tony learned some skills of disclosure, checking and inquiry, but used them in ways that served forms of reasoning that were still basically unilateral. His reasoning was still oriented to the protection of self and others from the discomfort involved in holding each other accountable.

Reasons for Limited Impact

It is important to explore the reasons for this limitation, because lessons could be learned about how to make future interventions more effective. One possibility is that Tony's encapsulation of his new skills reflected an inevitable learning sequence, and that as he experienced the consequences of more open communication he would change his beliefs about the risks involved. This argument essentially claims that the adoption of the strategies of critical dialogue will over time lead to the adoption of its underlying values and reasoning processes. The argument is flawed, however, because Tony's current ways of reasoning led him to make private unilateral judgements about when the use of critical dialogue skills was inappropriate. These decisions in turn severely limited the opportunity he had to test whether or not the new skills would have the consequences he predicted. If one believes that a staff member is too fragile, sick or insecure to be challenged, then one is caught in a self-sealing trap; the belief cannot be tested.

A second reason for the limitation is that the intervention failed to focus in a sufficiently concentrated way upon those very reasoning processes. Tony had predicted that we would not "get behind his skin" and although we did at times disclose our understanding of the way his reasoning limited his learning, we failed to establish a shared and sustained focus on those processes. In the following extract the executive has been tackling the problem of how to acknowledge "big issues", even though time pressure prevents them from being immediately discussed. They discuss how both the issue and the lack of time can be publicly acknowledged by dropping what they called a "marker buoy". The researcher then warned:

> *VR*: Now if you do lots of those and you never actually make the time to resolve the bigger issue then you'll become distrusted because you're marking issues that you're not following up, so you've learned a communication skill from us that is actually . . .

> *Tony*: Never consummated.
>
> *VR*: Never consummated, right, so that will be mistrusted. . . . I'm suggesting that you do the marking, that you learn that skill, but that you do have to commit yourself to finding or making the time to deal with the big issues, and that if you don't do that all that we're doing now will be another sort of game [NGTR40, 12.7.88].

We failed to follow up this type of challenge, however, with detailed discussion with Tony of how we believed his reasoning limited his understanding and application of critical dialogue.

There is a third related reason for the limitation in Tony's learning. With the benefit of considerable hindsight, the researchers can now say something about the way limits in their own understanding of critical dialogue at that stage contributed to the limits in Tony's learning. The focus of the third learning seminar for both principals was raising and resolving difficult issues. Tony wanted to rehearse a conversation he intended to have with one of his Heads of Department who had failed to use agreed school procedures in reporting to parents. Tony wanted to learn why the mistake had been made, satisfy himself that it would not happen again, and do so in a way that produced commitment to the decision rather than grudging compliance. We performed a role-play in which we repeatedly rehearsed parts of the conversation, guided throughout by a theory of participation that we thought would overcome the HOD's reluctance to discuss the matter and lead to greater mutual understanding. The theory involved facilitating participation by careful listening to concerns, by attempting to publicly test those concerns, and by disclosing and checking one's own views. We assumed that if the HOD did not become more committed to the problem-solving process, then there was a problem with the way we had engaged him and that this could be rectified by recycling through the process to detect what had gone wrong. Tony demonstrated the skills of disclosure, illustration, checking and facilitation, but the HOD remained disenchanted, convinced that the new report system was a retrograde step and that further meetings with either his own department or the principal were a waste of time. Tony declared at the subsequent session that the theory had been "up against the wall" and he was right, because despite six or seven attempts it had not produced the consequences we predicted [NG/WCTR3, 3.5.88].

We had misunderstood the theory in two ways. First, we had treated the outcome of internal commitment as something that resulted from a process of dialogue and, second, we had seen it as our responsibility for generating that commitment "in" the HOD. These are misunderstandings, because while commitment should be enhanced by critical

TABLE 7.6. *Analysis of Researchers' Unsuccessful Intervention*

Researchers chose to	*HOD chose to*	*Researchers ignored*
• focus on the way the reports might be perceived by parents. • focus on the way consistent reporting procedures could be achieved next time. • take responsibility for HOD's lack of engagement in this discussion.	• test his students differently from his colleagues. • report the results in a way that was inconsistent with agreed procedures. • unilaterally evaluate his error as "not mattering". • opt out of further discussion of reporting procedures.	• HOD's self-sealing logic, e.g. his unilateral evaluation of his own error as "not mattering". • HOD's rejection of his colleagues through his unilateral decision to test and report student progress differently, and to not engage in further discussion. • HOD's responsibility to alter "habits" that undermine agreed procedures.

dialogue skills, it cannot generate commitment in people unwilling to engage in that dialogue or unwilling to recognise the interdependence that such dialogue presupposes. While the HOD rejects a notion of interdependence, facilitative discussion about his reports to parents is a waste of time. The problem we should have focused on was the way the

HOD sought to disconnect himself from his colleagues and to justify his disconnection through self-sealing reasoning processes. Table 7.6 summarises the actual and suggested alternative focus of our intervention (C. Argyris, personal communication, 25.9.88).

The third column of the table addresses issues of authority in participatory processes by asserting the interdependence of organisational members and their responsibility to check their own judgements and reasoning processes against those of colleagues. Without an acceptance of this type of authority, individuals can thumb their noses at participatory processes and the decisions that result from them. Leaders must call to account those who unilaterally reject such processes, or they will not learn how to correct errors in those processes, and nor will the individuals learn about the responsibilities that come with collegial and participatory forms of organisation. Such forms of organisation are undermined as much by an educational culture which uses notions of collegiality and autonomy as defensive strategies to avoid accountability, as by the more commonly discussed problem of autocratic leadership.

What can we conclude about the impact of the learning seminars on Tony's style? By learning how to convey his views openly, Tony had reduced his earlier tendency to ease-in to difficult conversations. He had also become very skilled at detecting the taken-for-granted assumptions in his own and others' speech. His use of these skills was still limited, however, by his continued unilateral protection of himself and others from difficult issues, as illustrated in Table 7.5, and by his and *our* failure to come to grips with the processes of accountability implicit within critical dialogue. Tony had learned the communication skills necessary to avoid the autocracy that he feared, by encouraging the participation of his staff and the expression of diversity. However, the researchers had not helped him to learn how critical dialogue would enable him and his staff to weld from this diversity a unified and co-ordinated school programme to which all would be mutually and openly accountable.

Addressing the School-based Problems

The evidence presented in the previous section suggests that our attempts to alter the theory Tony brought to the resolution of tough problems had thus far been relatively unsuccessful. We did not have an opportunity to work through these limitations with him, because after three years as principal of Northern Grammar Tony accepted a position with the Ministry of Education. A series of follow-up interviews with the staff we had initially talked to showed that the issues that we had raised in our initial analysis remained.

The more positive results we achieved with Carol meant that we could go on to address the question of the impact of our intervention on the way Carol and her senior staff tackled problems in the professional development programme. The steps taken by the researchers, and increasingly by Carol and her executive staff, to address the problems in the programme are described and evaluated in the next four sections. The first describes the learning seminars for the executive staff, the second, the process of redesigning the professional development programme, and the third and fourth draw on the follow-up interviews to report staff perceptions of changes in Carol and in the new programme.

Redesigning Professional Development at Western College

We had not originally intended to formally involve the executive staff in the research project, as it seemed to us that the benefits of critical dialogue could be mediated to others through the changed style of the principal. Carol advocated that we alter this strategy and we agreed, believing that the intervention would be faster, more open and more supportive if others were involved in the work of developing and sustaining the new culture. Carol also argued that the executive held many of the same assumptions about professional development as she had held, and she wanted our help with checking these beliefs and involving the executive in the problem-solving process. In addition, the executive wanted to be involved because the research that their principal devoted so much time to was something of a mystery, and if any benefits were to be gained, they wanted to be part of them.

These arguments led us to replicate the same research cycle with the executive that we had employed with Carol, except over a considerably shorter time frame. Over a series of five meetings, our analysis of the problems of impact and staff resistance in the professional development programme was debated alongside a discussion of the relevance of critical dialogue to the resolution of these problems. Once again, both the validity of our analysis and the relevance of our alternative theory were tested as the executive debated our diagnosis and recommendations for change.

The result of those meetings was a request from the executive that they have more time to think about the material and to discuss it among themselves before reporting back to the researchers. Their discussion led to an agreement that Carol should write a paper for staff, in which she expressed her reasons for wanting to redesign the professional development programme, and a request to the researchers to run a one-day workshop for the executive on critical dialogue. Carol's paper went through five drafts as she responded to feedback from the researchers, her executive and the professional development committee. In it she acknowledged that the issue of appraisal had not been squarely faced in the previous scheme and that she had consulted staff about the details of the programme but not about its overall shape.

> *Carol*: In hindsight, I think it has been unwise to avoid the use of the term *staff appraisal* in relation to the [professional development] consultation cycle. I do not intend to avoid the responsibility I have as principal to appraise the performance of staff. Appraisal involves making judgments and this should have been clarified.

Carol now wanted to involve all the staff in designing a programme which faced up to issues of accountability and development in a collaborative way. The next six months saw a round of consultation with various staff groups as they debated a definition of appraisal, and the roles and responsibilities of those involved. Alongside the formulation of policy, staff discussed their need for training in the skills of giving and receiving difficult feedback and of problem-solving. The executive staff, with the help of the researchers, ran two one-day courses for those staff responsible for appraising others. While participants did not become proficient in a day, the courses gave them a shared espoused theory and a basis for monitoring and evaluating the implementation of the new scheme.

The process of change was long and at times stressful for Carol. In responding to an earlier draft of this chapter, Carol highlighted some of the difficulties she experienced. One was the necessity to confront staff who believed that consultation about the revised programme was a waste of time because "they would have to do it anyway". They expected the principal to make a unilateral decision and "get on with it". Another stress was the practical difficulties involved in facilitating such a lengthy and complex change process. The existing meeting structures at Western College and Carol's organisational skill contributed greatly to the completion of the process.

Changes to the Professional Development Programme

The success of the intervention is evaluated against two different criteria. One involves examination of the new professional development policies and procedures and asks whether they were designed to make an impact on teaching and learning processes and to surface rather than hide difficult issues. The other criterion involves staff attitudes towards the new scheme, and asks whether the extensive consultation process has resulted in a scheme to which staff feel more committed. It would also be desirable to assess whether the programme as practised reflected these qualities, but our follow-up work was cut short, as Carol left to take up a position in a polytechnic as director of a new educational management centre before the first cycle of PDC processes under the new scheme got underway.

In June 1989 Carol produced a draft discussion document on staff appraisal. The document includes a rationale, statement of purpose, guidelines and a conclusion before illustrating some of the more practical features of how the new programme might operate (Appendix A). The document illustrates many of the features of critical dialogue. It clearly states the purpose as being the improvement of classroom and management skills through the identification of strengths and weaknesses and the targeting of professional development activities to the latter. It also stresses the necessity to make judgements about the worth of the work that is being appraised.

> There is no doubt that critical review and judgement must be employed in the process of appraisal, nor can those responsible for the performance of others avoid the professional responsibility of appraising (Appendix A, p. 2).

The guidelines encompass three principles of openness, reciprocity and positiveness, designed to prevent the appraisal processes being used to unilaterally control the work of the appraisee. The validity of judgements is to be tested through disclosure and discussion; power differentials between appraiser and appraisee which may prevent open dialogue are modified by the use of reciprocal appraisal, and the appraisal process ends with agreement about how performance can be improved, regardless of its present level.

Although this document does represent a successful translation of the values of critical dialogue into a school policy, it does not reflect an entirely independent test of Carol's learning. The researchers commented on drafts and helped with the formulation of the guiding principles. A more stringent test of the effect of critical dialogue on the culture of professional development at Western College is provided by an analysis of the final appraisal policy which had no input from the researchers. This final policy was written by a committee of three, chosen by the whole staff to reflect a range of staff reactions to Carol's policy draft. The paper clearly acknowledges that staff appraisal is an integral part of staff development, and "that without staff appraisal, professional development does not effectively target individual and organisational needs" (Appendix B).

The paper also incorporated the three principles of openness, reciprocity and positive focus. It differed from the draft policy, however, in that Carol's inclusion of "staff strengths and weaknesses" was replaced by the more euphemistic description of "strengths and needs". This may be a minor difference in wording. On the other hand, it may reflect a continuing preoccupation with threat avoidance and a failure to recognise that critical dialogue provides ways of facing up to problems and weaknesses without demeaning the professionalism of teachers.

Changes in Carol's Leadership Style

Follow-up interviews were conducted with eight of the ten staff who were interviewed at the outset of the project. The interviews focused on staff perceptions of the management style of the principal and whether there had been any change in this perception over the course of the two years. In addition, we wished to know whether the reintroduction of professional development in the school was viewed by staff as having been handled in a more open and consultative manner than previously.

Staff members who were closest to the principal in the school's management structure (i.e. the other members of the executive) perceived the greatest change in her style. All three saw the training as having had a major impact on the way Carol managed her interactions with them and with other staff. They saw her as much more likely to encourage people to contribute and much more able to engage people in genuinely consultative processes.

> *Exec. member*: I think she's much—her style now is much more likely to encourage people to contribute and to talk ... the consultative thing is much more there, and trying to get people to tell things as they see it rather than trying to always be convincing other people of how she sees it [WCTR65, 20.11.89].

At the same time, some staff also saw these changes in Carol as incomplete, noting that at times she still had a tendency to talk over objections, particularly if challenged.

> *Exec. member*: I think—I think right back two years ago I suggested Carol—if you got into a debate with Carol and she felt threatened by the debate she became very verbal. And because she's a very gifted person, verbally, it becomes impossible to negotiate with her. Now I've used the word negotiate and I don't think I'd ever have used the word negotiate two years ago. But I think that's still potentially a difficulty. Umm, I say it because I saw it just the other day, in a very difficult situation to handle ... for her ... I just felt that she ... reacted too quickly, like she has always done, but then also on the other hand, particularly perhaps in the executive staff communication, she's much more relaxed [WCTR63, 22.11.89].

The members of the executive had received the most training in critical dialogue, and had themselves taken responsibility for teaching it to senior staff responsible for staff appraisal. They strongly supported this approach because they saw it as highly relevant to their work, and they were delighted with the benefits their own training had had on their ability to communicate with each other.

Given that the executive staff were knowledgeable about critical dialogue, they were likely to have been more sensitive than other staff to any changes in Carol's leadership style. The reactions of the other five staff who were interviewed seemed to largely depend on the frequency and nature of their prior contacts with her. All felt that they now enjoyed an easy relationship with her in which they were able to raise important issues. Two of them had always found her to be open and influenceable. One other had found her to have become more open in her role as principal than when she was deputy-principal, and the other two saw her as having become more open over the course of the research.

In summary then, three executive members and two staff members noticed changes in her openness over the course of the research, and felt that she had become more influenceable.

> *Staff*: "... she was, let's say authoritarian before, but she knew what she wanted and you wouldn't change her mind previously. Now I think she is willing to listen to people's views but occasionally she shows what she wants" [WCTR62, 20.11.89].

The other three staff members saw her as being open at the start of the project, in that they could not recall having had any problematic interactions with her.

Staff Attitudes Towards the New Scheme

Despite the fact that every school was now required to take responsibility for the professional development and appraisal of their staff (Department of Education, 1988), staff did not feel that the forms of professional development and appraisal advocated by Carol had been in any way imposed by her. They reported feeling that they had had a real opportunity to influence the design of the programme so that it met their needs as best as possible. As one staff member noted:

> ... the P.D. discussion was guided by Carol but she was open about her agenda and staff perceived her as influenceable ... [WCTR62, 20.11.89].

The more open and consultative style that staff commented on with regard to Carol's style is also reflected in their reactions to the way the appraisal policy was renegotiated and rewritten. All of the interviewed staff felt the renegotiation had been conducted in an open and consultative manner.

> *Staff*: She has been very concerned that—um—people who wish to be consulted are consulted. Very concerned about that. And very open about it. It was a matter that could have been dealt with in a totally different way. It could have been presented as a proposal that came straight out of the mouth of management and presented to the staff—say "there, what do you think of that?" And it wasn't, it was aired first of all. And then once it was aired, there was lots of ideas that came up. Management went away and produced a proposal—a draft proposal. And then we went through the process of consulting and discussing and changing. And I think it was a very good thing to do. Air it first of all in a very open manner [WCTR60, 20.11.89].

Staff were also asked how committed they felt to the new policy. The members of the executive all noted that they had always felt committed to the appraisal process as a management responsibility, but one noted an increased commitment to show that the new policy works. All staff but one stated that they felt more committed to carrying out the new appraisal policy, with one noting his continued commitment to appraisal. Many of those interviewed raised doubts about how well their colleagues understood the new policy despite the consultative way it had been negotiated.

> I have always felt committed to (appraisal) but I am still not sure that it is fully understood by all staff [WCQN, 15.2.90].

They wondered whether the doubts and nervousness of some of their colleagues might be an inevitable part of the process of change in a large organisation.

> People are beginning to accept as an integral part of their teaching the idea of appraisal. There is still an element of nervousness and this will continue until staff have experienced an appraisal and see it as non-threatening [WCQN, 15.2.90].

In this last comment there is also the possibility that some staff still equate "non-threatening" with threat avoidance. Consolidation of the new appraisal culture will require repeated checking and challenging of how staff understand such language, and its implications for improving the quality of teaching and administration at Western College.

In summary, staff do seem to believe that there have been some real changes in Carol's style, in the way the appraisal policy was renegotiated and in the final form of the document. They indicate either a continuing or increased commitment to its successful implementation while at the same time acknowledging that there is some way to go before they will fully believe that real changes have occurred for the school as a whole.

Carol and Tony Reflect on the Process

Given the considerable commitment and courage that this intervention called for from the two principals, it is appropriate that they be given the last word. Carol experienced our feedback paper in which we argued that she herself was part of the problem as the most traumatic aspect of the whole process. While it had caused her considerable anguish, it had triggered in her a much deeper level of understanding about what learning critical dialogue involved. Before the paper, she had treated the process as analogous to learning a new vocabulary:

> Learning and applying critical dialogue required me to adopt a new and somewhat artificial vocabulary which was at odds with my usual language of interpersonal management. I found the language learning difficult, asked for and received help, but felt, however, that I was still demonstrating effective practical skill in communication and problem-solving situations with staff (Cardno, 1990).

After the feedback paper, which triggered a strong emotional reaction, she saw how critical dialogue, in addition to offering a "new vocabulary", challenged some of the values that informed her existing practice.

> My beliefs and values had never before been challenged or held up to critical review . . . my response was emotive to begin with. . . . A defensive reaction was inevitable. I believed I could justify aspects of my style as being practical, necessary, efficient, and based on expectations which others held of me. These views were mutually explored and on reflection I had to agree that, however painful, the analysis was accurate. This was no easy process of acknowledgment, and involved a form of catharsis in which my feelings were transferred to paper. . . . I found that I was able to identify unilateral values of avoidance of unpleasantness and wanting to be right! For me, this was a turning point in understanding the theory that would help me own and confront tough issues. So, from this point on, I was motivated by the fact that this skill training was beginning to make sense; that I was capable of accepting and exploring my own practice as it appeared to others, and that I could perhaps salvage both my self-esteem and my programme, by sticking with the research commitment (Cardno, 1990).

By this stage Carol was seeing a clear connection between the skills she was learning and her ability to resolve the problems we had identified.

> Embarking on the [problem-solving] phase as a key actor was a daunting experience. Firstly, it involved a huge commitment of time; secondly it raised anxieties about my ability to cope and the commitment level of other actors on whom I would depend; thirdly, it required a shift, not only in confronting technique but also in my assumptions about fundamental issues of teacher autonomy and performance evaluation (Cardno, 1990).

Tony was more positive about the impact of the workshops than we were. He believed that while he had a considerable way to go, he had learned skills that enabled him to be both more effective and less anxious in his dealings with others. During a follow-up interview, we spent some time discussing his perception of his own learning:

> *Tony*: When we started talking you said, "Had you succeeded in teaching me critical dialogue?" Well, I'd have to say I think I know what they are, and I feel I have demonstrated them and can demonstrate them and am using them in this job. . . . I feel that my whole way of operating changed. I have often said that you have to place yourself on a continuum somewhere. I moved toward critical dialogue—how far, I'm probably the last person to judge, but certainly in my beliefs about myself I did . . .
>
> . . . I must have the feeling that occasions, events have happened in which the use and application of critical dialogue theory has caused things to go better than they might otherwise have done. I've had that experience again and again and I'm prepared to assert that so have other people who have dealt with me . . . [NGTR61, 20.3.90].

Tony was aware of limitations in his own ability to behave consistently with the theory, limitations which he put down to his need for more time and, above all, to the complexity of trying to use the theory in the midst of the factionalism and win–lose politics of school life. He saw the theory as standing apart from and slightly naive in its treatment of these issues. In attempting to understand the difficulty of applying critical dialogue in his school, he pondered:

> . . . straight personality differences must be a huge part of it and, again, I didn't ever find the theory taking those things sufficiently into account. In other words the theory operates in a very pure . . . supposedly pure environment, and that is why in the early parts of things I really wanted you to know what the politics of the school

were. Because that I think is a large part of the answer. . . . But I don't want to float that around as if I'm not part of the problem . . .

. . . just what people want, what barrow they're pushing, who's jockeying for position and who's shafting the other in the race for promotions and so on, is all part of the workings of a school, which the theory doesn't sufficiently deal with [NGTR61, 20.3.90].

While Tony has a strong sense of his own development, his reflections also capture the limitations in our own teaching, particularly our over-emphasis on skills, which is again apparent in the way Tony understands our initial question. He is not confident that the theory speaks to the politics of organisational life, and so sees the two as sitting uneasily alongside one another. Our view was that critical dialogue takes the win–lose politics of school life into account through its emphasis on the way unilateral patterns of interpersonal and organisational life constantly interfere with genuine inquiry. If the culture is to be modified, these patterns of unilateral control must be constantly confronted and undermined, and doing so may provoke conflict and upset. It was Tony's reluctance to confront, and our failure to teach him its importance, that left him feeling that critical dialogue was naive in its treatment of school politics.

Carol too was aware of the way the politics of school life posed formidable challenges to her ability to behave consistently with her newly espoused leadership style. Her reaction to these difficulties, however, was somewhat different from that of Tony. She made conscious choices, which she then checked with the researchers, about the contexts in which she felt that the politics were too tough, and in which she, therefore, would not be able to maintain her open stance. We confirmed her wish to take things a step at a time, to manage her stress levels, and to involve us in helping her to tackle progressively more difficult situations. Carol also began to sense that critical dialogue had implications for patterns of school organisation and structure as well as for interpersonal interaction. Having perceived what it meant at the micro-level, she began to consider its implications for the more macro features of school organisation.

Carol: See, what you have switched from as I see it, is that we started out with democracy in a one-to-one situation. We're now talking about managing the school in a democratic way which is a very much wider issue, and in a way I wasn't ready for the wider view. I'm still really working with individuals. I've always, I suppose, part of my value or belief is that you must have all those things like participative structure and all the mechanisms in place, but I was still—and that's all very structural and very distant—

but in one-to-one situations what was very revealing was how little democracy there really was. Now you're making me worry about the larger democratization of the school in terms of what I thought was there . . . [NG/WCTR3, 3.6.88].

Summary

PBM suggests that educational problems are resolved through the critique of relevant theories of action and the provision or development of an alternative. Our critique was debated at length with the principals in a series of feedback meetings designed to test the accuracy and completeness of our analyses and the appropriateness of the alternative practices that they implied. The principals' initially defensive reactions did not stop them, in time, from raising important issues which contributed to the revision of each analysis. Both Tony and Carol found the analyses sufficiently compelling, and the alternative sufficiently attractive, to agree to proceed to an intervention phase.

The first objective of this phase was to teach the theory and practice of critical dialogue, so that the principals themselves would become more skilled at inquiring into and resolving the problems they faced. The second, assuming that the first was achieved, involved helping them to apply this approach to the resolution of the school-based issues which were the focus of each case study. The first objective was addressed by teaching the theory and practice of critical dialogue while simultaneously investigating the ways in which deeply held unilateral values and strategies interfered with this learning. The teaching and learning process emphasised the importance of focusing on actual practice rather than on self-reports, of raising and replying to objections to the alternative, and of debating and practising the theories in the context of real school issues.

As a result of these workshops, close colleagues reported Carol to be more open, willing to listen and consultative. These perceptions were confirmed by evidence from meetings which Carol led to discuss the redesign of the professional development programme. While Tony became quite skilled in some of the strategies of critical dialogue, he was still highly selective in how he used them, avoiding, in particular, situations in which he anticipated threat or embarrassment to himself or others. Tony had grafted these skills onto unilateral ways of reasoning, partly because we had failed to focus sufficiently intensively on the theories behind Tony's threat avoidance. We did not have a chance to work through these limitations with him because he left to take up a new position in the Ministry of Education. A series of follow-up interviews showed that, contrary to his self-perceptions, Tony's staff perceived no change in his style, or in the problems they had previously reported in the school's transition to more democratic management.

In the second phase of our work with Carol, the problems of the professional development programme were addressed by including the executive in the training, and by a lengthy process of consultation and debate with the whole staff about the problems of the previous programme and how to correct them. This resulted in a programme which made explicit the connection between staff appraisal and development, and which was based on principles of openness, reciprocal appraisal and positive benefit. Most of the staff we interviewed reported that they had been fully consulted about the programme, and that they felt committed to it. The chapter concluded with Carol's and Tony's reflections on the intervention process.

References

Argyris, C. (1970). *Intervention theory and method.* Reading, MA: Addison-Wesley.

Argyris, C. (1976). *Increasing leadership effectiveness.* New York: Wiley.

Argyris, C. (1985). *Strategy, change and defensive routines.* Boston: Pitman.

Argyris, C. & Schön, D. (1978). *Organizational learning: A theory of action perspective.* Reading, MA: Addison-Wesley.

Block, P. (1981). *Flawless consulting: A guide to getting your expertise used.* San Diego: University Associates.

Cardno, C. (1990). Collaborative research: A subject's perspective. Unpublished manuscript.

Department of Education (1988). *Tomorrow's schools: The reform of educational administration in New Zealand.* Wellington: Government Printer.

8

Reflections on Intervention

In reflecting on the intervention phase of the two cases, I was guided by two broad questions: "How adequate was our intervention?" and "How does it compare to other more traditional approaches to educational change?" The first question is addressed by applying the criteria of theoretical adequacy to the theory that informed the interventions. At times, a distinction is made between our espoused theory of intervention and our theory-in-use, because some inadequacy was attributable, particularly at Northern Grammar, to our failure to live up to our own espousals. The consideration of theoretical adequacy introduces several potential objections to the change process associated with PBM, and these are discussed in the second major section of this chapter. Along the way, an answer is offered about why the intervention was relatively more successful at Western College than at Northern Grammar. The second question, comparing the theory of change incorporated in PBM with that advocated by other educational writers, is answered in the context of Fullan's (1991) recent volume on educational change.

More on Theoretical Adequacy

In judging explanatory accuracy in Chapter 6, the focus was on our explanations of the problems we had been asked to investigate at Western College and Northern Grammar. In judging effectiveness, coherence and improvability, the focus shifts from our explanation of the problems to the way we attempted to solve them. As explained in Chapter 2, there is not always a direct link between an explanatory theory and a theory of intervention, because the former may not yield actionable solution steps, or the latter may be constructed independently of any knowledge of problem causation. In the two cases discussed in Part II, however, the theory of the problem feeds directly into the theory of the solution.

Western College

At Western College the solution comprised two interrelated components. The first involved challenging the dichotomy that staff had set up between accountability and collegiality and reframing it so that they could tackle tough issues in ways that served values of learning and problem-solving rather than values of unilateral protection and unilateral control. The second interrelated component involved teaching the skills and values of critical dialogue, because we believed that they were necessary for achieving the goals associated with the first. The adequacy of this solution is reviewed in terms of the specifics of our intervention at Western College and in terms of wider issues of its underlying logic and appropriateness.

Effectiveness

In evaluating the intervention described in Chapter 7, I suggested that the principal, and to a lesser extent staff, had learned how to address issues of accountability and quality in ways that would enable them to conduct a more effective staff development and appraisal process. These positive changes are not sufficient to establish the effectiveness of our theory, however, because it may not have been our intervention that caused staff to rethink and redesign their programme. Attributing the change to our intervention requires elimination of competing explanations. The most obvious competitor is the introduction, during the time of our project, of a national policy of school self-management (Department of Education, 1988). In effect, the policy gave responsibility for staff development and appraisal to the principal, and while Western College had already had considerable experience with the former, it had made less progress with the latter. The new policy signalled an official expectation that staff be appraised, and that school personnel, rather than external authorities, be responsible for doing so. The result was that the debate within schools shifted from whether there would be programmes of staff appraisal to the principles and practices that should comprise them.

I would argue that this policy change acted as a facilitating condition for our intervention, but that without the latter, the problems in the programme would not have been solved. First, nothing in the new policy addressed the dilemmas that were already being experienced in the Western College programme. The expectations were higher, but the policy gave little recognition to, or assistance with, the difficulties involved in meeting them. Second, without the intervention, the policy change could have exacerbated the problems, because as accountability pressures on school leaders increased, the costs of avoiding tough issues would have become unacceptably high. Without skills in critical dialogue, it is likely that Carol and senior staff would have resorted to more unilaterally controlling strategies, thereby increasing the resistance of some staff and jeopardising the learning and problem-solving potential that comes with a mutually educative appraisal process. In short, although the policy changes made it more important and urgent that the problems in the programme be addressed, the history of the problem and an analysis of senior staff's skills suggest that those changes alone would have been insufficient for their resolution. This leaves open the reverse question of whether the intervention would have been successful without the policy change. My answer is inevitably speculative, but while I believe that the change made little difference to Carol's motivation to address the problems in the programme, it probably meant that staff accepted the need for appraisal more quickly than they would have otherwise.

The conclusion that the effectiveness of the intervention was enhanced by the policy change is supported by research on the conditions that promote successful educational change. Fullan (1991) summarises the relevant research this way: "Successful change projects always include elements of both pressure and support. Pressure without support leads to resistance and alienation; support without pressure leads to drift or waste of resources" (p. 91). I would add to Fullan's point that when pressure sets a requirement to address a problem rather than to implement a particular solution, it is more likely to be productive, because it does not pre-empt the problem-solving processes needed to make effective local changes.

Coherence

The coherence criterion of theory appraisal sets the researcher's theory in the wider context of all our knowledge about schools, educational administration, teaching and learning and any other relevant domain. It asks whether, despite its apparent success, the theory is criticisable because it is inconsistent with what we have reason to believe about these other domains, and with what would be required to solve problems that arise within them.

The application of the coherence criterion can be illustrated through discussion of the frequent claim that critical dialogue is threatening and raises teacher stress to unacceptable levels. Some might argue that our solution to the problems in the professional development programme is incoherent, because it is inconsistent with what we have reason to believe is interpersonally effective. Critical dialogue may solve some problems in the programme, but at the cost, for example, of unacceptably high levels of teacher stress. It is important to remember, in testing the validity of this claim, that the coherence criterion demands consistency with what we have reason to believe and not with what we believe. The belief that critical dialogue raises stress levels is not reasonable when it is based on untested attributions about others' emotional states, or on a misunderstanding of such dialogue as involving forthright, closed-minded expressions of one's opinion. It may be true, however, that critical dialogue, even when correctly understood and practised, does increase teacher stress in some cases. The question to be then asked is whether the theory that was effective in resolving the original problems of the programme is compatible with a solution to this subsequent problem of teacher stress. If stress is understood as a result of a mismatch between job demands and teacher capacities, analysis and resolution of this problem may require discussion of the same type of difficult issue as did the problems in the professional development programme. The theory that underpins the intervention has the potential to resolve both problems; the theory that underlies the objection is incompatible with what is required to solve any problem whose discussion has a potential to threaten or embarrass one or more participants. My conclusion, therefore, is that the theory underlying the intervention is more coherent than the one that underlies this frequent objection to critical dialogue.

Improvability

The processes of critical dialogue are the means by which on-line theory improvement is accomplished in PBM. Errors are detected and corrected, and theories made more comprehensive and effective, through making them explicit and thus exposing their claims to public testing. In addition, the deliberate search for competing concepts, explanations and solutions sets up competition between rival theories in a way that strengthens the eventual theory choice.

There are a number of important objections to the claimed association between critical dialogue and theory improvement. They centre on the question of whether or not such dialogue presupposes conditions which are generally not present in organisational and interpersonal life. The first concerns the relative expertise of participants in the dialogue. If critical dialogue requires roughly equal expertise among the participants, then it is ruled out in many cases, including that of our intervention at Western College. Given that the researchers had far greater expertise than Carol in reconstructing and evaluating theories of action, one could question the claim that she could contribute meaningfully to a dialogue about the relative merits of our theories of professional development.

The second objection concerns the claim that critical dialogue results in warranted agreement about the detection and correction of error.

Does the theory of critical dialogue overstate the possibility of reaching a consensus about the interpretation of events and about what action to take? Is it based on an overly consensual model of human interaction? Since these objections apply equally to both cases and deserve detailed consideration, they will be considered under a separate heading, after a brief discussion of the adequacy of our theory of intervention at Northern Grammar.

Northern Grammar

At Northern Grammar, our problem analysis suggested that Tony's failure to be open about important aspects of his vision, and his difficulty in confronting some of the practices that were incompatible with that vision, jeopardised his desire to develop a staff who were guided, not by the principal's decrees, but by a shared and workable philosophy for the school. This analysis suggested that a solution lay, in part, with Tony learning how to negotiate his vision with staff, not only at the level of espoused theory, but also at the level of the concrete realities of its implementation. Our intervention at Northern Grammar, therefore, was designed to teach Tony the skills and values of critical dialogue so that greater mutual understanding and commitment to a shared and realisable vision of the school would be the eventual result.

Effectiveness

The ineffectiveness of our intervention at Northern Grammar leaves us trying to sort out how much can be attributed to faulty implementation, and how much to weaknesses in the theory itself. Although in Chapter 7 I attributed much of Tony's failure to change his theory-in-use to our unskilled intervention, it is possible that the theory would not have been effective no matter how skilfully employed. It is possible that the theory of critical dialogue underestimates the emotional and cultural defences involved in threat avoidance, and hence underestimates the barriers to open inquiry into difficult issues. If this is the case, it would help us explain why the intervention was more successful at Western College than at Northern Grammar. Our analysis of the style of the two principals suggested that Carol was far more willing than Tony to confront colleagues face-to-face. While Tony was frequently unilaterally protective, Carol was more likely to be unilaterally controlling, as seen in her declarations about what was wrong, and about what needed to be done to put it right. If the theory of critical dialogue underestimates emotional barriers to openness, this would have been more detrimental to our work with Tony, because he held the value of threat avoidance far more strongly than did Carol.

Criticisms of Critical Dialogue

We now have three possible objections to the theory of critical dialogue, centred on inequalities of expertise, the resolution of disagreements and emotional barriers to openness. Since critical dialogue is a key to the success of intervention in PBM, the above objections need serious discussion. If such dialogue is only possible under highly unusual conditions of organisational and interpersonal life, then it is criticisable as idealist and utopian, unable to offer practical help in the resolution of educational problems. The discussion is illustrated with excerpts from conversations between the researchers and principals which are relevant to the three objections.

Inequality in Expertise

If bilateral management of the key tasks of PBM (Table 3.1) requires roughly equal competence in those tasks on the part of all participants, then bilaterality is rarely possible, because at least at the beginning of an intervention process the researcher's relevant expertise is probably greater than that of practitioners. The question is, then, do differences between participants in relevant knowledge and expertise prevent the exercise of bilateral control? This inequality did not prevent the principals in our two cases from raising numerous questions about the accuracy of our problem analyses and about the appropriateness of our suggested interventions. Raising objections, however, is not the same as competently defending them, so perhaps it is here that inequalities in expertise preclude bilateral control. The argument is flawed, however, because it rests on the mistaken assumption that inequalities in contribution to the content of a decision automatically translate into inequalities in interpersonal control over that decision. A researcher with expertise in the critique of theories of action, for example, will make a greater contribution to the content of that critique than those with less, but this contribution does not, in principle, preclude equality of interpersonal influence over the final evaluative decision. A practitioner who has made little contribution to the critique can still disagree, raise objections and, if not convinced by the arguments and evidence, decide against its acceptance. Once these two types of inequality are separated out, it becomes clear that inequalities in contribution to the content of a decision do not necessitate unilateral control over decision processes.

Objectors might still argue that the distinction is fine in principle, but that in practice, those who have less expertise will tend to agree with those who have more, because they will feel unable to articulate the basis for their doubts and hesitations. Critical dialogue recognises this possibility in its emphasis on facilitation as a necessary condition for bilaterality in such contexts. The job of researchers is not only to gain reactions to their theory, but also to help those who have less expertise to articulate the theories that are a source of their doubts and hesitations. In this way, inequalities in expertise are less likely to translate into inequality in control over decision-making.

A second defence against the objection that inequality in expertise precludes bilaterality appeals to the educative qualities of critical dialogue which simultaneously recognises such inequalities while working to reduce them. If differences in relevant competence are openly discussed, then all parties can work to reduce them, where desired, as efficiently as possible. This may involve, paradoxically, the use of educative processes that, taken out of context, look quite unilateral. At times in our workshops, we used a number of direct teaching strategies such as mini-lectures, structured exercises and coaching techniques. I do not believe these processes were unilateral, however, because they were the mutually agreed means by which we tried to reach shared goals. In addition, we frequently checked the usefulness of these processes. Sometimes unsolicited feedback on this approach would also reassure us of the appropriateness of this style, as when Tony commented at the end of one of our sessions, "Good one, I feel like I've been taught something."

In summary, inequality of expertise does not preclude bilateral control in PBM, because a distinction is made between equality of contribution to the content of a decision and equality of interpersonal control over the making of a decision. The facilitation skills of critical dialogue help to ensure that inequality in the former does not automatically translate into inequality in the latter.

The Resolution of Disagreements

The claim that critical dialogue promotes theory improvement implies that these processes produce agreement between participants about the relative merits of various theoretical claims, resulting in a shared understanding of the problem and of how to resolve it. Given that many educational problems arise, or are at least not resolvable, because key players disagree over the identification, the analysis or the resolution of the problem, what are the features of critical dialogue that promote such shared meaning? The first is the commitment to focus on rather than bypass the disagreement, and to do so in the context of a problem which motivates the participants to make progress. The second is the value base of critical dialogue which treats disagreement as a resource for theory improvement, rather than as an obstacle to the persuasion of others. The third feature is the interpersonal strategies which are designed to test the adequacy of the beliefs which are the basis of the disagreement.

Disagreements about what occurred can be resolved by checking one's perceptions with others and by accurately recalling and reporting events. It sounds simple, but in practice, values of unilateral control and winning frequently lead us to report events in ways that are designed to serve our own interests. If we view our boss as autocratic, we are likely to report on what happened at a staff meeting in ways that omit the times others' opinions were asked for and was influenced by them. When the boss disagrees with such descriptions of practice, the issue may be resolved by retrieving the examples which were initially omitted by those who sought to portray him or her as an autocrat.

People may agree on the data, but still disagree on how to interpret it. Some may argue that despite the instances of consultation, the boss is still an autocrat, because the instances of consultation were relatively trivial compared to the occasions of non-consultation. The basis of the disagreement is now theoretical differences about what counts as autocratic leadership, rather than differences in reports about what occurred. PBM requires that we attempt to resolve the differences by applying the criteria of theoretical adequacy. Which theory explains more of the boss's behaviour and has more predictive power? Which theory is more coherent with our understanding of related concepts of leadership and interpersonal influence? In addition, the criterion of improvability suggests that we ask how the holders of the competing views would test their beliefs. In other words, what evidence or argument would count in each case as a disconfirmation of the theory in question? For example, what type of consultation would the boss have to engage in for those who believed that he/she was an autocrat to change their opinion? If their theory does not allow for such possibilities, then it is self-sealing and less adequate than its competitor.

There will be cases where disagreements will remain despite conscientious employment of these moves, and other cases where practical constraints like time, goodwill and cognitive capacity prevent their use. As indicated in Table 3.1, critical dialogue recognises this possibility, and requires that such cases be managed in a bilateral rather than unilateral fashion. Our feedback to Tony led to such a disagreement. Much of our analysis was based on widely shared staff perceptions of his style. For every such example we produced, however, Tony either produced alternative more positive interpretations of the same examples, or positive counter-examples. If true, his responses suggested that our interview data were incomplete at best, or systematically biased at worst. We shifted our focus to a collaborative meta-level conversation that acknowledged the disagreement and the need for a decision about how to proceed.

VR: Well, I suppose our sense from our feedback session and then our talking over the phone was that, we may all be heading towards Rome, but . . . we're on different roads.

Tony: Yes.

VR: That's what the feedback and your reaction to it suggests to me. And so one thing that we'd like to do today is to see whether we can agree about what road we're going to be on.

Tony: Mmmmmm. [. . .]

We then discussed whether we needed to resolve the differences between Tony's and some of his staff's perceptions of aspects of the school's functioning. Tony began to describe his view that the task forces he had set up to look at school management were working well. The researchers then referred back to their interview data and suggested that there was still a puzzle.

MA: In a way that leaves me feeling a bit stuck, because there's quite a—it feels like a big discrepancy between what you're feeling and what we're picking up.

Tony: Mmmm.

MA: From out there.

Tony: Well, that was the puzzle that you were referring to, wasn't it?

VR: Yes, on the phone, yes.

We then began to discuss how to resolve the puzzle that we had identified. Gathering more evidence from staff was unlikely to resolve the impasse, because the same differences were likely to re-emerge. It was more important to understand the processes that had led the staff and principal to hold such widely disparate perceptions of Tony's style and of his attempts to implement his vision of the school. We suggested that instead of continuing to discuss our interview data, we look more directly at how Tony and the staff gave each other feedback, with a view to evaluating their skill at testing and checking the attributions they made about each other.

VR: And so, we're suggesting that instead of us giving you more feedback, we facilitate a process whereby you get it from the people that matter.

Tony: Yeah, that sounds to me like hands on sort of stuff.

VR: Yes. Right.

Tony: I like that [NGTR30, 3.3.88].

We could not agree on the data, but we could agree on a process for continuing our collaboration without sacrificing its critical thrust. After another round of interviews with staff, several more discussions with Tony, and three more drafts of the analysis, we began to understand the gulf between Tony's perceptions and those of his staff. Tony's "resistance" had pushed us to develop a theory which was now more comprehensive and more compelling. The research problem was no longer one of deciding whose perception of the principal was right, but one of explaining why they were so disparate despite an espousal, on Tony's part at least, of a staff united in their commitment to a shared vision of the school.

Critical dialogue does not guarantee resolution of all disagreements. Given the underdetermination of theory by evidence (Phillips, 1987, pp. 11–16), then it is likely that no amount of rational debate and interpersonal skill will resolve some of them. The claim of critical dialogue is to reduce the probability of such disagreements and the possibility that unresolved disagreements will become destructive. Resolutions occur largely through unpicking the bases of the disagreement and searching for shared logical, normative or empirical standards against which to test the competing views. When agreement is still not possible, critical dialogue requires a shift to a meta-level conversation about how the disagreement should be managed. Tony and the researchers agreed in the above example that despite their disagreement, they were still committed to the project. Our past work together and our future plans provided a context of goodwill that enabled us to work out how to proceed. In essence, we asked each other "What is the significance of this disagreement? Does it mean that we quit? Is there a way we can proceed with integrity despite it?" A bilateral relationship can sustain a few unresolved disagreements because people do not insist on immediate resolution, they focus on what is agreed to, and they trust that with time and continued openness, the disagreement will either be resolved, or take a form that opens up new options.

Emotional Barriers to Inquiry

When critical dialogue was first introduced in Chapter 3, it was contrasted with a model of interaction whose primary values were not inquiry and problem-solving, but unilateral control and the avoidance of unpleasantness. The contrast made it clear that inquiry into difficult issues required us to overcome a natural tendency to avoid anticipated emotional unpleasantness. As Argyris writes in a book on defensive routines in organisations: "... truth is a good idea when it is not embarrassing or threatening—the very conditions under which truth is especially needed" (Argyris, 1990, p. xiv). In the face of actual or predicted threat and embarrassment, people avoid inquiring into the source of the threat in order to protect themselves and others from emotional upset. They also cover up their avoidance, since to reveal it is itself embarrassing. Such defensive reasoning is strongly supported in Western culture by the meanings we give to such social virtues as caring, support and respect of others (Argyris, 1990, pp. 106–107). Since most of the intractable problems that face educational leaders pose the possibility of threat to one or more of those involved, one result of such reasoning is that these problems become undiscussable, except in a form that distances them from the people who may be responsible (Bridges, 1986). Another consequence is that the belief that they are undiscussable cannot itself be publicly tested.

The prior discussion of critical dialogue suggested several ways in which the conflict between inquiry and emotional threat could be overcome. First, inquiry which eliminates prejudgement and encourages reciprocal influence reduces defensiveness, and so makes exploration of difficult issues less stressful, without sacrificing learning and problem-solving. Second, bilateral rather than unilateral management of anticipated or actual emotionality implies that one tests rather than assumes the correctness of one's attributions about how others may react to the inquiry process. It also implies that trade-offs between emotionality and inquiry become discussable rather than privately managed by one party. Third, since critical dialogue involves appraisal of the relative adequacy of competing theories, it also requires us to understand and publicly evaluate the reasoning processes that sustain the emotional barriers to inquiry.

Given that the emotional barriers to critical dialogue were not successfully overcome in our work with Tony, the question arises as to whether the above steps are adequate to the task. While further empirical work may indicate that this is not the case, the researchers' failure to fully implement them at Northern Grammar makes it difficult to reach a conclusion. While we taught the executive staff the skills involved in "tackling tough issues", we did not critically evaluate the theories that sustained their fears of upsetting one another, and that prevented them from resolving problems in their team. We got closest to these issues in the extract reproduced in Chapter 7, where Tony explains his fears of being tough with his colleagues, and of becoming an autocrat if he ever overcame that fear (p. 147). This extract is rich in clues about Tony's theory of negative emotion, and rather than explore them further, the researchers rushed on to designing the following week's skill training session. If we had probed the significance of the distinction Tony made between confronting those he worked with and confronting others, the link he made between being more skilled at tackling tough issues and becoming an autocrat, and his fear of his own power, we could have inquired into his theory of negative affect in a way that made it possible for us all to systematically compare it with the values of critical dialogue.

Without these steps, we can say more about the limitations of our skills as interventionists than about the limitations of critical dialogue. If we had completed them, and Tony had still been unable to incorporate inquiry values into his theory-in-use (he had always espoused these values), then we would have evidence that some people cannot set aside their deep-seated defensive barriers to inquiry, even though they are aware of them and of their consequences. Perhaps some people will continue to choose their own and others' short-run happiness above inquiry in situations where they perceive the two to be in conflict, even though they have an accurate understanding of the possible consequences and an opportunity to learn an alternative. Critical dialogue, in other words, even if properly employed, cannot guarantee that defensive barriers to inquiry will be overcome, because, in the end, participants may still choose to give a higher value to avoiding emotional discomfort than to inquiring into its source. If many practitioners were to make such a choice, however, despite skilled intervention, then the adequacy of critical dialogue as a path to the resolution of educational problems would be in doubt.

It is interesting to take a reflexive turn at this point, and to consider the possibility that emotional barriers to inquiry were preventing the researchers themselves from inquiring into the barriers they perceived as operating in the Northern Grammar executive team. Our feedback to Tony had been very challenging, and at one stage our disagreements had threatened to derail the project. At stake was the completion of a research contract and two years' investment of time and energy. The situation was made even more difficult by the fact that Tony's deputy had threatened to discontinue her involvement, if the workshops we conducted with the executive probed any further into team relationships. The researchers raised the issue with the team, but in the absence of Tony's support and the refusal of the deputy to discuss it any further we decided to "play it safe" and pursue less threatening skill training issues, without publicly considering the consequences of this choice.

Interventionists more skilled than myself have written about the way they have fallen into the same trap of not challenging client pressure to bypass these emotional issues. Mangham (1988), for example, in evaluating his work with an executive team, describes how a lack of persistence with such issues may well be regretted later.

> My attempts were met by anger and aggression, by blame and contempt. I desisted because I did not wish to embarrass or further discomfort [the leader]. The cost became obvious later. Feelings will be aroused in all such events; they should always be addressed, never avoided (p. 147).

While agreeing with Mangham's general sentiments, I would want to qualify them somewhat. First, the intervention processes involved in PBM do not require that all emotional issues be addressed. The exploration of emotionality is only undertaken in PBM in the context of its effects on inquiry into and resolution of important work-related issues. This qualification provides one of the distinctions between PBM and the T-group and encounter group movements of the sixties, in which emotionality was sometimes valued for its own sake. Research on these movements has shown how such an emphasis can be irrelevant to work-related effectiveness (Kaplan, 1979). A second qualification on Mangham's injunction that feelings must be addressed may also be required depending on his precise meaning. If he means that the feelings that lead people to avoid dealing with some issues should be addressed, then I would agree, because those feelings may prevent progress in resolving the focus problem, and so a decision not to deal with them should be made in full knowledge of its implications for the success of the intervention. If he means that the interventionist should unilaterally decide to deal with the issue itself, then I disagree, because such intervention violates the value base of critical dialogue, and could, depending on the power of the interventionist, inflict psychological harm on some participants (Lieberman, Yalom & Miles, 1973).

In summary, the adjudication of competing theories, when conducted in an interpersonal context, requires a form of dialogue that challenges if not directly contradicts many people's current interpersonal practice of avoiding emotional issues. Unless this practice is altered, the problem-solving and inquiry processes of PBM will be continually misunderstood, avoided or incorrectly applied. Strong emotional barriers to the inquiry processes of PBM and the researchers' inability to overcome them were largely responsible for the relative failure of our intervention at Northern Grammar. If more skilled intervention than ours still results in frequent rejection of these processes, then considerable revision of the theory of intervention that informs PBM will be required.

Another Perspective on Change

One way to judge the distinctiveness and merit of the PBM approach to change is to compare it with that offered by other writers on intervention. The recent volume by Michael Fullan (1991) is particularly apposite, because it is a highly influential work which develops a theory of planned educational change, from a comprehensive review of empirical research since the sixties. Fullan explains his purposes as follows:

> I have attempted in this book to distil from these experiences the most powerful lessons about how to cope with and influence educational change. In compiling the best of theory and practice, my goal has been to explain why change processes work as they do and to identify what would have to be done to improve our success rate (Fullan, 1991, p. xi).

The quote suggests that Fullan's purposes include the development of a descriptive theory of the way change is conducted, as well as a normative theory of how it ought to be conducted in order to "improve our success rate". An educational change, according to Fullan, requires change at three levels of practice: in the use of new or revised materials, in the use of new teaching approaches, and in the alteration of relevant beliefs (p. 37).

Fullan's emphasis on change provides an immediate contrast with PBM, for the latter is not concerned with change *per se*, but with the improvement of educational practice, understood as the resolution of problems. Before tracing the implications of this difference, it must be acknowledged that Fullan repeatedly cautions his readers not to take the desirability of change for granted, and to make a distinction between change and progress. Despite these cautions, the distinction does little theoretical work in Fullan's treatise. A theory of change requires a technically sound set of guidelines for altering the status quo. A theory of progress requires, in addition, normative guidelines for judging the desirability of the processes and purposes of change. Fullan hints at such guidelines when he suggests that progress can be judged by whether or not the innovation "helps schools accomplish their goals more effectively by replacing some structures, programs and/or practices with better ones" (p. 15). Later on, however, Fullan shifts to a subjectivist position on the judgement of progress, answering the question "How do we know if a particular change is valuable, and who decides?" with the statement "The short answer is that a change is good depending on one's values, whether or not it gets implemented, and with what consequences" (p. 45). Given that Fullan has not tackled the complexity of elaborating and defending these various criteria, including resolving the potential contradictions between them, he offers a technical theory of change, and not a theory of educational progress. In PBM, a normative framework is developed, centred on the promotion of inquiry and learning, against which both the processes and purposes of change can be evaluated.

Fullan's approach to the process of change is also quite different from that of PBM. While change is conceived in PBM as a process of problem-analysis and resolution, it is conceived by Fullan as a three-stage process of initiating, implementing and institutionalising an innovation.

> In simple terms, someone or some group, for whatever reasons, initiates or promotes a certain program or direction of change. The direction of change, which may be more or less defined at the early stages, moves to a phase of attempted use (implementation), which can be more or less effective in that use may or may not be accomplished. Continuation is an extension of the implementation phase in that the new program is sustained beyond the first year or two (or whatever time frame is chosen) (p. 48).

Fullan's theory of change comprises discussion of those factors which determine the way each stage is conducted, with particular attention to those that are predictive of successful completion. For example, Fullan begins his discussion of initiation by listing eight factors which influence the decision to proceed with a change. These include the availability of high quality innovations, the access of decision-makers to information about these innovations, the degree to which administrators, teachers and external change agents advocate for particular changes, the degree to which the community, or sections of it, is mobilised to promote or resist particular reforms, and finally, whether school districts are motivated by a problem-solving or bureaucratic orientation to change (pp. 50–61). The discussion then shifts to the processes associated with successful initiation which are defined as start-ups "that have a better chance of mobilizing people and resources toward the implementation of desired change" (p. 62). Fullan suggests that initiations are more likely to be successful when they are relevant (perceived as offering a clear way of meeting a high priority need), and when the school is ready to initiate change. Readiness describes the institution's practical and conceptual capacity to change, including the possession by staff of the requisite understandings and skills. Finally, resource availability influences whether a decision to initiate change leads to real change in practice.

There are several important implications of the difference between the way change is conceptualised in PBM and in Fullan's theory (Table 8.1). Most of them reflect the greater discipline imposed on the change process when it is guided by a theory or, more accurately, a sequence of theories, about the nature of the problem and its resolution. The first implication of the difference concerns what counts as a significant

TABLE 8.1. *A Comparison of the Theory of Change of PBM and Fullan (1991)*

	PBM	*Fullan*
Definition of educational change	Altered practices sufficient to resolve a problem.	Altered beliefs, teaching approach and use of materials.
Scope and type of change	Determined by the specifics of the problem situation.	Change at all three levels required in all cases.
Desirability of change	Judged by multiple criteria of solution adequacy.	Normative theory not developed.
Process of change	Process of problem analysis and resolution embedded in a critical dialogue between all stakeholders.	Progress through three stages of adoption, implementation and institutionalisation of an innovation.
Initiation of change	Decision by one or more stakeholders that a problem is worth investigating.	Decision to adopt or develop an innovation.
Motivational thrust	The desire to solve a problem experienced as unacceptable by those involved.	Various: including need, career incentives, resource availability, innovation advocacy.

educational change. Fullan insists that any significant educational change must occur at the three levels of resources, teacher strategy and belief. In PBM, the degree and scope of change required is determined by the problem analysis, so a significant change may not involve all three levels. Imagine a school district where first-grade teachers had an outstanding theoretical and practical competence in the teaching of beginning reading, but the children were reading at below their expected level because the district could not afford to buy appropriate texts. Resolution of this hypothetical problem only requires change at the level of resources, not at the level, in addition, of belief and teaching strategy. The significance of the change should rest on the importance of the problem it addresses and its success in doing so, not on its scope or type.

The second, and perhaps most important, difference between the two theories of change concerns the relative emphases given to critical analysis of existing practice. Fullan's theory begins with the decision to initiate change, that is, adopt or develop an innovation. In PBM, change starts with the identification and analysis of a problem of practice. This process is important in developing a shared understanding of the likely causes of the problem, and of what counts as a solution, including whether or not the relevant factors are or can be brought under the control of those involved. A shared understanding of the problem enables participants to plan a change process that is regulated by that knowledge, and that can be monitored against shared criteria of what counts as progress.

Fullan might object that analysis of current practice does play a part in his theory, because the relevance of the innovation, that is its match to a high priority need, is critical to the success of his initiation phase. My argument would still stand, however, because the identification of a need is not the same as an analysis of a problem. For example, the staff at Western College identified a need to reduce teacher stress, and sought to meet it through professional development workshops on relaxation and stress management techniques. They did not conduct a problem analysis which would involve critical inquiry into the assumptions that led them to specify stress as the problem for which relaxation was the solution, or part of it. The result of such inquiry could be, for example, recognition that the teachers' stress was related to the design of their jobs, and could not be overcome without major restructuring of teacher timetables and responsibilities. The problem analysis process not only provides a deeper understanding of the problem itself, but also indicates whether aspects of the culture will, without intervention, prevent its resolution. In short, the inquiry process needs to include the question that Sarason (1990, p. 13) asks in his book about the failure of school reform, "Why should this reform attempt be any more effective than the last?"

Fullan and PBM are similar in their emphasis on the importance of the way practitioners understand their current practice and the proposed changes. The former sees the main reason for the failure of school reform as neglect of the meanings that practitioners give to proposed changes.

> ... developers or decision-makers went through a process of acquiring their meaning of the new curriculum. But when it was presented to teachers, there was no provision made for allowing them to work out the meaning of the changes for themselves (p. 112).

The process of critical dialogue in PBM not only recognises the importance of meaning, but in addition, the importance of convergence on a meaning which is sufficiently shared to allow co-ordinated action to be taken towards problem resolution. Fullan, unlike PBM, provides no mechanism by which differences in meaning or, in the language of

PBM, differences in theory can be adjudicated. While recognising, in common with PBM, the social dimension of educational change, his theory does not include a normatively based process for consensually co-ordinating that social dimension.

Finally, a change process that is problem focused, as in PBM, offers significant motivational advantages over one that is not. The stress and disruption that change inevitably brings is more likely to be accepted when it holds the promise of overcoming a situation that those involved have agreed is unacceptable. Without such a perspective, the motivational equation is unbalanced, because the costs of change are so much more salient that the promise of future benefit. This comparison between the approach to change of PBM and of Fullan has demonstrated the implications of focusing change on problem analysis and resolution rather than on the adoption, initiation and continuation of innovations. A problem focus, together with a normative theory of solution adequacy, provides a basis for establishing criteria for the planning and evaluation of change, a motivational impetus, and a basis for regulating the social learning that is essential for co-ordinated action.

Summary

Critical dialogue was both the researchers' means of intervention into the practitioners' problems and the intermediate goal of that intervention, since it was hoped that the principals themselves would subsequently use this form of dialogue to involve their own staff in the problem-solving process. The adequacy of the theory and practice of critical dialogue was judged on criteria of effectiveness, coherence and improvability. On effectiveness, the mixed success of the intervention raises questions about either the theory of critical dialogue, the way it was practised, or both. It was argued that our limited success at Northern Grammar was due to the substantial emotional barriers to inquiry that were encountered, and the researchers' failure to inquire into and alter the theories which participants held about how to be effective in the face of actual or anticipated emotionality. If these barriers continue to prove insurmountable in future research, despite more skilled interventions, then the effectiveness of the theory itself must be questioned.

On the criterion of coherence, it was argued that while critical dialogue frequently reveals more problems, these are more likely, in turn, to be resolved by such dialogue than by the predominant unilateral approaches to problem-solving. On the criterion of improvability, a number of objections to the claim that critical dialogue fosters the detection and correction of error were discussed. Objections about the impossibility of engaging in such dialogue across differences in expertise were answered by showing that there is no necessary connection between inequalities in contribution to the content of a decision and inequalities in interpersonal control over that decision. Critical dialogue incorporates steps designed to reduce the probability of such a connection and to reduce such inequalities over the course of an intervention.

The objection that critical dialogue exaggerates the possibility of resolving disagreements is frequently based on inappropriate generalisation from unilaterally controlling forms of interaction. The inquiry-oriented values and skills of critical dialogue encourage identification of the basis of disagreement and public testing of its implications for shared criteria of theoretical adequacy. In addition, critical dialogue recognises that not all disagreements are resolvable, and incorporates steps suggesting how remaining disagreements can be bilaterally rather than unilaterally managed.

The comparison of PBM with Fullan's theory of change highlighted the implications and advantages of a problem-based rather than an innovation-based approach to educational change. A problem focus motivates effort, co-ordinates action, and provides shared criteria against which progress can be monitored. While PBM and Fullan's approach share a focus on the meanings practitioners bring to a change process, only PBM incorporates a theory for collaborative discovery, evaluation and, if necessary, revision of those meanings.

References

Argyris, C. (1990). *Overcoming organizational defenses.* Boston: Allyn and Bacon.

Department of Education (1988). *Tomorrow's schools: The reform of education administration in New Zealand.* Wellington: Government Printer.

Fullan, M. (1991). *The new meaning of educational change.* New York: Teachers College Press.

Kaplan, R. E. (1979). The utility of maintaining work relationships openly: An experimental study. *Journal of Applied Behavioral Science*, **15** (1), 41–59.

Lieberman, M. A., Yalom, I. D. & Miles, M. (1973). *Encounter groups: First facts.* New York: Basic Books.

Mangham, I. (1988). *Effecting organizational change.* Oxford: Basil Blackwell.

Phillips, D. (1987). *Philosophy, science and social inquiry.* Oxford: Pergamon Press.

Sarason, S. (1990). *The predictable failure of educational reform.* San Francisco: Jossey-Bass.

PART III

Comparing Problem-based Methodology

9

Empiricist Research and Problem Resolution

Two types of argument are central to the claim that problem-based methodology is particularly suited to understanding and resolving educational problems. The first type, which was the subject of Part I, is concerned to show how PBM is well suited to these purposes through its inclusion of such characteristics as a constraint inclusion account of problems and a theory of critical dialogue. The second type involves demonstrating that existing educational research methodologies do not already incorporate these characteristics, or that they do so to only a limited extent. The comparison is essential to establishing the appropriateness of PBM to the practical purposes of problem understanding and resolution. If existing research traditions already incorporate many of the features of PBM, then there is likely to be nothing distinctive about the latter's contribution to the understanding and resolution of educational problems. In addition, this second comparative argument requires showing that the absence in other methodologies of those features which are distinctive of PBM somehow jeopardises their practical contribution. It is this second argument which is the subject of the third part of this book.

Before launching into the comparative process, it is worth restating some of the broad features of PBM that are most germane to its comparison with existing educational and social science methodologies. Fundamental to PBM is its purpose of understanding educational problems in order to resolve them. Achieving this purpose requires theories that not only meet science's traditional concern with explaining the world, but that also tell us how to change it. The four standards of theoretical adequacy associated with PBM, first described in Chapter 2, reflect the complexity of these purposes because they go beyond traditional concerns about empirical adequacy to include, in addition, normative adequacy and effectiveness. (In distinguishing empirical and normative adequacy I am not taking sides on the highly contentious fact-value distinction, only reminding readers that PBM is concerned with both.) To omit standards of normative adequacy is to ignore an important requirement of practice and to engage in the sort of technical reasoning that some social theorists think is at the heart of many of our social and educational problems (Habermas, 1984). To omit a concern with effectiveness is to neglect the qualities of a theory, over and above its empirical accuracy, that make change more or less likely. It is extremely difficult, given the breadth of PBM's concept of theoretical adequacy, to make straightforward links between this methodology and a particular epistemology. Despite these difficulties, and at the risk of oversimplification, I shall describe PBM as a non-positivist empiricism, whose particular features can best be understood by revisiting the four criteria of theoretical adequacy described in Chapter 2, and then comparing them with the characteristics of more positivist empiricisms.

Empirical adequacy, as was evident in the two cases, is critical to the analysis and resolution of educational problems. Theories are to be preferred which incorporate accurate causal accounts of the phenomena which they seek to explain. Explanatory accuracy is promoted through a process of disconfirmation; that is, through subjecting favoured accounts to tests generated by plausible alternative theories of a problem. PBM admits both objective and subjective states as causal; attributions about subjective states are subject to the same sort of test as any other causal account. For example, an attribution about an emotional state is tested against an alternative by examining its relative ability to predict the actions we associate with such states. Empirical adequacy is also involved in the criteria of effectiveness and coherence. Our attempts to resolve educational problems will not be effective if they are based on inaccurate causal accounts or faulty theories of change. The effectiveness criterion, however, also assesses those non-empirical qualities which make theories differentially successful in resolving problems. Causal accounts which show how particular agents can act to change problematic situations are to be preferred to those which do not; theories of change which require modest rather than heroic effort are to be preferred because they cohere better with our knowledge about the motivational requirements of successful change (Weick, 1984). The coherence criterion assesses the extent to which a theory about a particular problem and how to solve it is consistent with all our knowledge of the world. This criterion is critically important to PBM because it prevents a kind of *ad hoc* problem-solving that might relieve the immediate practical concern, at the cost of our ability to resolve other current or future problems. This criterion is indirectly empirical because what counts as our best theory of the world has already been determined, at least in part, by relevant empirical tests.

I turn now to the more difficult question of how to justify the inclusion of the normative standard of improvability in PBM. The difficulty arises because it is not obvious how values, in the sense of the affirmation of particular values, can be justified by appeal to experience. When I say that I value theories and practices that foster improvability, just what grounds do I have for that valuing? Answering this question involves discussion of how an empiricist methodology can admit values in general, and how the particular value of improvability can be justified. Kaplan (1964) suggests that the most appropriate grounding for values is some sort of naturalism, by which he means that "the ground of values consists, with suitable and important qualifications, in the satisfaction of human wants, needs, desires, interests and the like" (p. 389). Since matters of fact arise in showing that the value in question does indeed promote relevant wants, needs and interests, we may be mistaken in our value claims and thus the issue of their objectivity can be legitimately raised. In other words, the researcher can ask, as was done repeatedly in the case studies, whether actors have good grounds for holding the values that they do. The particular value of improvability is defended in PBM by arguing that the ability to learn and to solve problems that arise in the pursuit of our wants and interests is basic to their achievement; that without this value we jeopardise the achievement of our other values, through failing to learn over time how to lead our lives in ways that can fulfil them.

Research Traditions or Paradigms?

Given the recent burgeoning of methodological approaches to social and educational research, some grouping is necessary in order to make the comparison process manageable. The broad categories of empiricist, interpretive and critical were chosen because they are increasingly recognised as distinctive research traditions, even though there are numerous varieties within each, and there is controversy over the strength of the boundaries between them (Bredo & Feinberg, 1982; Carr & Kemmis, 1986). The position taken will be, that while PBM has some features in common with each of the traditions, none of them has the particular combination of features that is found in PBM, and that its absence reduces the practical contribution of the approach. For example, while the emphasis on causal reasoning in all empiricist research is compatible with PBM, positivist versions of empiricism place strictures on inquiry into the causal role of practitioner reasoning and on the evaluation of values. These strictures severely limit the contribution of this methodology to the improvement of practice. Both PBM and interpretive research give a central place to practitioners' understandings, but the latter usually lacks a problem focus and a normative perspective from which to develop an alternative practice. On the surface, PBM and critical research have most in common, particularly in their normatively based critique, but it is argued that the theory of interests that informs much critical theory, and its neglect of effectiveness, limit its practical contribution. After a short discussion on the question of paradigms, each of the three traditions will be briefly described, a comparison made with PBM, and an example of research conducted within the tradition analysed. The research examples included in Part III were chosen as illustrative of deliberate and thoughtful attempts to do research in a way that made a difference to practice.

To claim that PBM incorporates key features of empiricist, interpretive and critical approaches to research is to accept that those approaches are, in the main, able to be integrated. Readers who are familiar with the paradigms debate in educational and social research will recognise that such integration entails a rejection of paradigm theory (Lincoln & Guba, 1985; Smith & Heshusius, 1986). This theory treats empiricist, interpretive and critical approaches to inquiry, and their variants, not just as different methodologies, but as different epistemologies with different assumptions about the nature of explanation, truth and the justification of knowledge claims. These differing assumptions render knowledge claims generated within one paradigm immune to criticism from another, since each paradigm incorporates distinctive standards of adequacy. If paradigm theory is correct, the three research traditions are incommensurable, and cannot be integrated in the way that is claimed in PBM. If, on the other hand, paradigm theory, in its strong incommensurable sense, is false, then the current diversity of approaches to educational research can be seen as a common pool of resources for the development of a methodology appropriate to the goals of problem understanding and resolution.

Although a full-scale defence of the integration, via refutation of paradigms theory, is beyond the scope of the current work, a summary of some of the arguments, based on the work of Walker and Evers (1988), is provided, so that readers know where to position PBM within this paradigms debate. They begin their attack on paradigm theory and their argument for the epistemological unity of educational research with the following claim: "In beginning our defence of the unity thesis we note that in philosophy, and philosophy of science in particular, P-theory [paradigm theory] is widely regarded as false" (p. 32). Their critique of paradigm theory is based on the self-referential argument that P-theory or any other epistemology "should be knowable on its own account of knowledge" (p. 34). What then does P-theory say about how we know that educational research is partitioned into paradigms? If the claim is itself part of a paradigm, then the standards that apply to its justification (whether they be observation of the activities of researchers, giving of reasons, or understanding how researchers understand their own activity) are those that apply in that paradigm. Those already committed to the paradigm will be persuaded; those who are not will be unconvinced. If the claim that educational research is partitioned into paradigms lies outside any particular paradigm, or is applicable to all of them, there must be some standards that are applicable across paradigms, and once this is acknowledged, the argument about the incommensurability of the paradigms begins to break down.

The point of all this, in terms of the forthcoming comparison of PBM with existing traditions of social research, is to establish that PBM is neither a new research paradigm nor a version of an existing one. Given my dispute with the existing paradigm discourse, I have no wish to reinforce it by locating PBM within one of these alleged entities. Having said this, I have no problem with recognising that there are different traditions and approaches to educational and social research and that PBM draws on those features of these traditions and adds some new ones that suit its particular purposes of problem understanding and resolution.

Empiricist Approaches to Educational Research

A major difficulty in comparing PBM with empiricist approaches is characterising the latter in a way that does not stereotype the approach for the sake of exaggerating the difference between it and PBM. In order to avoid such stereotyping and to make room for the considerable variety of approaches within this rubric, a broad definition of empiricism is offered. Lacey (1986, p. 61) defines it in *A dictionary of philosophy* as:

> Any of a variety of views to the effect that either our concepts or our knowledge are, wholly or partly, based on experience through the senses and introspection. . . . Extreme empiricists may confine our knowledge to statements about sense data, plus perhaps analytic statements. Less extreme empiricists say that such statements must form the basis on which all our other knowledge is erected.

The definition captures the central role of experience in the justification of our knowledge claims, while also indicating the wide spectrum of views about the precise role that experience plays. Although elaboration of the numerous varieties of empiricism is beyond the scope of this volume, the distinction between positivist and non-positivist empiricism is important because the degree of similarity with PBM is largely dependent on which account is taken as the point of comparison. I previously described PBM as a non-positivist empiricism, and this claim can now be explained further by comparing it with positivist versions. While positivism as an epistemology has long been shown in philosophy of science to be false, it is worth briefly reviewing its main features because some of its methodological tenets retain a strong hold over the practice of empirical research and, coincidentally, limit its contribution to the resolution of educational problems.

Although the origins of logical positivism can be traced much further back, it was at its height in the late 1930s, and is associated with an interdisciplinary group of scholars known as the Vienna Circle (Phillips, 1987, pp. 36–45 and 94–101; O'Hear, 1989, pp. 106–110). One of its key objectives was to rid philosophy and the study of human affairs of the dogmatism and metaphysical speculation that characterised contemporary scholarship in these areas. This was done by arguing that the methods of the natural sciences, with their emphasis on objective empirical inquiry, were the only appropriate methodology for the investigation of social life, and that subject matter that could not meet these strictures had no place within the new order. The verifiability principle of meaning, one of the main planks of logical positivism, was a powerful tool for excluding what positivists saw as unworthy knowledge claims. Phillips defines the verifiability principle as:

> ... something is meaningful if and only if it is verified empirically (i.e. directly, or if charitable, indirectly by observation via the senses), or is a tautology of mathematics or logic (Phillips, 1987, p. 39).

This principle served to exclude metaphysics and all theorising about unobservable abstract and moral issues from scientific inquiry. If we cannot know such ideas through empirical experience, and such experience is the only source of meaning (apart from analytic statements), then such inquiry will produce meaningless knowledge claims. For positivists, therefore, the meaning of a statement is the mode of verification, and the mode is that of confirmation through sense impressions (Kaplan, 1964, p. 36). Sense data provided a secure foundation for our knowledge because they were thought to yield direct experience of the world, uncontaminated by interpretation or prior theory. The verifiability principle of meaning has been thoroughly discredited because it led to various absurdities such as doubt over the status of scientific laws and the verifiability principle itself. The emphasis on verifiability remains strong in some empiricist research, however, in respect of the *justification* rather than the *meaning* of knowledge claims. Abstract theorising and explanation is still shunned in positivistically inclined empiricist research, not because it is meaningless, but because it is not amenable to direct empirical test.

Implications for Inquiry into Values and Subjective Experience

From the point of view of PBM, an epistemology that requires knowledge claims to be tested by direct observation poses problems because much of the subject matter with which it deals cannot be directly observed. For example, the values and understandings of practitioners which make up part of their theories of action cannot be directly observed, yet they play a central role in the explanation of problems of practice. Neither can the normative standard of theoretical adequacy (improvability) be defended, because the desirability of this or any other value cannot be directly apprehended. If positivist empiricism is correct, then these central tenets of PBM are in doubt.

Fortunately, PBM does not have to abandon its concern with the subjective and with values, because the strictures which positivist epistemologies would place on them are widely regarded to be indefensible. The now familiar arguments about the inseparability of theory and observation undermine the privilege accorded knowledge claims that can be verified through direct observation (Phillips, 1986; Evers & Lakomski, 1991). Humans do not directly perceive other persons, cats, snow or any other apparently concrete phenomena. These perceptions are the result of complex cognitive processing of patterns of retinal stimulation in terms of prior learning about the discrimination, identification and labelling of these patterns. In short, there is no sharp distinction between interpreted and uninterpreted experience, since even knowledge which is based on apparently direct observation is interpreted through prior theory about the significance of what we see. Once the sharp line between theory and observation is broken in this way, then observational evidence loses its privileged place; investigation of subjective states and understandings is no different in kind from investigation of objective phenomena; both involve theorising about the available evidence, though in the former case it is admitted that the nature and interpretation of the relevant evidence might be more contentious than in the latter. If theory and interpretation intrude into the investigation of both subjective and objective experience, then the positivist argument against inclusion of the subjective loses its force. Description and evaluation of theories of action become a legitimate focus of investigation in both PBM and non-positivist empiricism.

The same argument applies with respect to the inclusion of values in PBM. They have been avoided in positivist empiricism because there is no obvious way in which their truth or falsity can be established.

Empiricist researchers, therefore, can describe the values and ethical stance of their subjects, and investigate their antecedents and consequents, but cannot incorporate normative recommendations within their theories. The researcher's role, therefore, is to develop theories which predict the consequences of future choices, or evaluate the outcomes of past ones. They should leave practitioners and policy-makers to debate the desirability of the various options. Evers and Lakomski (1991) describe the influence of this stricture on research in educational administration.

> Now there is no denying that administrators with limited knowledge of the consequences of alternative courses of action would find immensely useful any theory that delivered more accurate predictions of events. But in many cases people know what the options are even if they don't know the detail. What they really want guidance on is the right course of action. Administrative theories that disqualify themselves from addressing the value question have a theory/practice problem: theory fails to be relevant for a large part of administrative practice (p. 10).

There is obviously some conflict between these strictures on normative claims and PBM. Precisely what the conflict is requires some careful analysis of the ways values enter this methodology. First, many of the constraints that practitioners set in the process of problem-solving are value constraints; that is, implicit or explicit prescriptions about how they and others ought to act. For example, in attempting to solve the problem of how to establish an effective professional development programme, Carol placed a high value on threat avoidance and on the promotion of quality teaching and learning. Now there is nothing in positivist empiricism which prevents the *description* of these values. As Michael Scriven puts it:

> ... they are obviously simply dispositional claims about an individual and hence open to objective investigation in the same way as are claims about the edibility of yesterday's lettuce (Scriven, 1972, p. 131).

Positivist empiricism is also compatible with the evaluation of value claims where matters of fact are relevant. For example, politicians who want to promote family life, while enacting economic policies that force both parents to seek paid employment, can be criticised for acting in ways that contradict their espoused values. Problems arise, however, when researchers wish to make the switch from internal to external critique, because the latter involves appraising the values of others rather than employing their evaluative standards, and thus the onus is on the researcher to defend the standard being employed. One such defence involves showing how the value in question leads to other already valued consequences, but positivist empiricists may argue that no matter how far one traced the empirical chain of consequences, one is always left with the question of how to justify the evaluation of the end point. Michael Scriven provides an answer which challenges the assumption that value claims are part of a hierarchy of knowledge, and that, therefore, some ultimate standard needs to be defended. He suggests that values should be treated like any other knowledge claim, in their need to be coherent with the rest of our global theory, responsive to empirical evidence and revisable.

> ... the most common objection to such views (that values can be included within the theories of social science) is raised by asking where the ultimate ends or "goods" come from. Perhaps the most satisfactory answer to this is to offer these alternatives: *either* there is no need to assume any ultimate values any more than we have to assume there was a first day in the history of the universe; *or* the ultimate (but not highest) values are the biological immutable needs of individual reasoning beings. The latter need not be thought of as necessarily valuable in themselves, only as desired, i.e. valued, in the first instance. That is, they provide an epistemological but not incorrigible starting point for the eventually self-corrective value system (p. 132).

Kaplan (1964) makes essentially the same point when he denies that every value judgement must come to rest upon an absolute end which is valued unconditionally. All ends are means to further ends, and thus the appraisal of values involves the assessment, in particular contexts, of the costs to be incurred and the benefits to be gained from the pursuit of any one of them. In this way, our whole set of values, rather than an ultimate value, is involved in the appraisal of any one of them "much as the whole of our knowledge is at stake in the test of any particular hypothesis" (p. 396).

In summary, the positivist empiricist can describe people's values, evaluate their consistency and trace their consequences, but not judge the worth of those consequences, because the basis of such judgements cannot be found in *a priori* reasoning, or in direct observation of the world. Acceptance of this stricture poses a problem for PBM, because it wants to argue for the value of improvability and hence for those theories and practices which promote it. The problem can be overcome, however, because the positivist privileging of observation is mistaken, and once this is recognised, then we can accept, following the arguments of Scriven, Kaplan and others, that the evaluation of values proceeds, in principle, along the same lines as for any other knowledge claim.

Implications for Theory Testing

In attacking the privilege accorded observations in positivist empiricism, my target is the assumption that such observations provide unambiguous theory-free and value-free evidence, not the role of empirical evidence *per se*. Unfortunately, as many philosophers of social science have pointed out (Phillips, 1987, p. 41; Kaplan, 1964), recognition of the serious problems of positivism has led some writers on social science methodology to a mistaken rejection of empiricist inquiry rather than to the adoption of a more open, holistic non-positivist empiricism. The importance of empirical constraints on theorising in PBM should be clear from the earlier discussion of criteria of empirical adequacy and from their constant employment in the two case studies.

The distinction between positivist and non-positivist forms of empiricism enables us to see that PBM is a version of the latter. Despite the emphasis put in PBM on empirical adequacy, however, readers may feel that the procedures by which this was established in the two cases in Part II were very different from those typically employed in non-positivist empirical research. I am referring to the quantification of variables and the specification of their relationships in hypotheses which are tested through collecting relevant evidence and subjecting it to a variety of statistical procedures. These procedures are generally not appropriate to PBM because they presuppose a degree of precision about a theory's constructs and the relationships between them that cannot be met during the process of developing a theory of an ill-structured problem. A heavy investment in, for example, a coding scheme or a multivariate analysis is premature when the data are still being explored to discover what meanings may be important to capture, and what relationships may be important to test. This is not to say that the logic of testing is not important to PBM; testing early hunches is, as we saw in Chapter 6, an integral part of developing a theory. The point is that the procedures of theory testing must be appropriate to the stage of theory development, and that premature use of traditional forms of hypothesis testing may result in a high degree of precision about uninterpretable phenomena or, as Hackman (1985) puts it, the constraints of a rigorous testing process may end up destroying the phenomena being studied.

Implications for the Social Relations of Inquiry

A second quite different argument for not employing these more typical empiricist testing processes in PBM is related to the social relations of inquiry. The procedures of rigorous quantification, experimental design and statistical analysis require specialist knowledge which is far more likely to be held by the researcher than the practitioner. The more technical and specialist the theory-testing process, the less accessible this phase of inquiry is to the practitioners with whom researchers need to develop a shared understanding of the problem. If special steps are taken, differential expertise may not necessarily be a barrier to collaboration, but when the expertise is highly technical, and is not essential to the competence of the practitioner, one questions the point of taking these steps so that practitioners can understand a highly technical theory-testing process. Carol did not need to know the precise percentage of her utterances that were unilaterally protective or controlling, or how they correlated with staff attitude, in order to understand and challenge the analysis presented in Figure 4.2. In fact, her engagement with that analysis may have been made even more difficult if it had been presented in this form. Instead, the rich examples, the preservation of the problem context and the attempt to present an analysis of the whole problem, rather than a fragment of it, made the analysis recognisable and challengeable by those involved.

The implications of rigorous testing procedures for the social relations of inquiry can be further explored by comparing the social relations of experimentation and of PBM. Experimental procedures incorporate tight controls which eliminate many, though not all, of the influences which confound the effect of the independent on the dependent variables. Such control is tightest in the laboratory experiment, with field experiments, field studies, surveys and case studies allowing for a progressively looser degree of control. For the purposes of theory testing a premium is therefore placed on experimentation. While considerable attention has been paid to the research procedures involved in experimentation, the social relations that usually accompany this form of research are typically unexamined (Walker & Evers, 1986; Gitlin, 1990). These will be briefly described, because I shall argue that they limit the extent to which experimental research can contribute to the improvement of practice.

The conduct of an experiment involves numerous decisions about the hypotheses to be tested, and about the procedures to be used in the introduction of the independent variable, in the observation of its effects on the dependent variable and in the way experimental control will be achieved. Since researchers have technical expertise about how to achieve experimental control, and that control sometimes requires that subjects are given minimal information about the experiment, researchers make these various decisions unilaterally. The achievement of experimental control, therefore, is accompanied by control of the subject by the researcher. The current emphasis on the "ethical" conduct of experiments, and on subjects' rights to give informed consent, to withdraw at any time and to be briefed and debriefed about the experiment, does not invalidate this claim. There is a difference between control and consent, and while these "ethical" procedures give subjects a clear choice about their involvement, that choice is about whether or not to give consent to being controlled.

Experimental control may jeopardise the practical contribution of research because it incorporates a theory of the subject that ignores the causal force of the meanings and motives that subjects bring to the experimental situation, and the degree of match of these forces to those which operate in the practical situation which the experiment seeks to inform (Walker & Evers, 1986). The experiment proceeds as if the only relevant influences on the subjects' behaviour are those specified by the researcher's theory, and as if those influences have the same meaning for the subject as they do for the researcher. Either the subject is seen as not having a relevant theory, or the experiment is designed to force or induce the subject to act within the framework of the researcher's theory. If the findings of experimental research are generated in conditions which force subjects to set aside their own theories, then they are not applicable in settings in which they are free to act in accord with their preferred view of the world. The social relations of experimental research make it very unlikely that the researcher will know when this is the case, and consequently, to what extent the findings of the research have any validity beyond the experimental setting. Walker and Evers put it this way:

> The point we would like to stress here is that social scientific laboratory experimenters *create* the object of their research: they arrange social reality so that it is amenable to their experimental techniques. This means not only that they cannot know whether their findings, e.g. about human learning, apply to learning in "real life" situations but that, ironically, they are having a massive "observer effect" on their object of study: indeed they are changing, through their domination of their S, the object [of study] (p. 384).

In summary, a comparison of PBM and empiricist forms of inquiry is impossible without drawing a distinction between positivist and non-positivist forms of empiricism. The former is significantly different from PBM, because its acceptance of observations as a secure foundation for knowledge leads to an inability to adjudicate competing claims about subjective experience and about values. Once the problems with such an epistemology are recognised, the considerable similarities between the epistemology of PBM and non-positivist empiricism become apparent. PBM's considerable emphasis on empirical adequacy makes sense because practitioners who act from inaccurate or seriously incomplete models of the world will simply not be as effective as those who model the world more accurately. PBM's emphasis on normative adequacy can be incorporated in a non-positivist empiricism because once the notion that observations provide an uninterpreted foundation for our knowledge is abandoned, value claims can be adjudicated, as Kaplan (1964) points out, in the same way as any other type of knowledge claim, by tracing their implications for and consistency with other undisputed knowledge claims.

Despite their epistemological similarities, however, there remain some important procedural differences between PBM and some of the research strategies associated with non-positivist empiricism. The first concerns the degree of formality and precision of the theory-testing procedures associated with each approach. It was argued that testing procedures that are demanding in terms of quantification and control are only appropriate when a theory is already well developed, and hence they are generally unsuited to the iterative process of theory development and testing that is typical of the analysis of problems in PBM. The second procedural difference concerns the social relations of inquiry that follow from the employment of more controlled theory-testing procedures. The exclusive focus on the researcher's theory, together with the suppression or ignoring of that of the practitioner, jeopardises the practical contribution of the research. The subjects' theories-in-use remain hidden, interacting in unknown ways with the research conditions, and varying in unknown ways from the theory they would employ in the context of practice. As a consequence, tightly controlled studies provide little opportunity for researchers and practitioners to critically evaluate the latters' theories and for researchers to pit their theories against those which will determine whether or not their findings will be applicable.

Illustrating the Comparison

The next section enriches the comparison between PBM and empiricist inquiry by discussing an example of a research project which attempted to do collaborative research in a way that improved practice. In the absence of detailed methodological let alone epistemological discussion by its authors, it is difficult to identify the extent to which the research was informed by a positivist or a non-positivist epistemology. For example, the choice to bypass the theories of action of the teachers may reflect either a positivist epistemology or a methodological choice about the most fruitful way of investigating this particular research question. Despite this epistemological unclarity, the example provides a rich source of comparison, at a procedural level, with the way problems are understood and resolved in PBM. In evaluating the study's methodology from the point of view of improving practice, I will focus on its treatment of problems, of theories of action and on its implicit theory of change.

During 1986 and 1987, the journal *Education and Urban Society* devoted two issues to discussion of recent examples of collaboration between educational researchers and school practitioners. One of the examples was the School–University Partnership for Educational Renewal (SUPER), a partnership between the Graduate School of Education at the University of California, Berkeley, and sixteen representative local schools, from kindergarten to college level. The journal's guest editor, the Dean of the Graduate School of Education, wrote: "The Project's central purpose is to develop models for implementing institutional change which will encourage lasting improvements in educational practice at school sites as well as within schools of education" (Editor, 1987, p. 363). There was a strong emphasis throughout the three-year project on both the processes and the structures required to foster collaboration between the two project partners. Up to eighteen months were spent negotiating entry to the various institutions, and the eventual research agenda was the product of canvassing the interests of both academics and teachers. Regular meetings were held, and newsletters and specially appointed liaison persons kept up to 2000 people informed of the project's activities. On the research side, monthly seminars attracted up to 175 people, and approved projects were conducted in schools which were interested and could provide access to classrooms and teachers. The various research projects conducted within the SUPER rubric are compatible with the empiricist tradition to varying degrees. The two described below are typical, however, because the first is concerned with precise identification of interactions between student and task variables and learning outcomes, and the third with the identification of observable regularities in the behaviour of school leaders.

> *The study skills project.* This project identifies and analyses those learning associated skills that are most effective at different levels of education, and whether the effectiveness of study skills varies with the characteristics of the student, the requirements of the teacher and/or the subject being taught.
>
> *The principal as instructional leader.* During the course of this study, a number of principals from SUPER schools were shadowed for 40 hours in an effort to gain an understanding of the varied demands on the time of a school site principal and to ascertain when and how educational leadership is possible within actual confines. A wide range of leadership styles and of school characteristics were examined (Gifford, 1986, p. 103).

Further description of the project will be given as needed in the methodological comparisons which follow.

Problem Focus

PBM starts from an initial problem statement and moves in repeated cycles of data collection and theory development until a theory is formulated which accounts for the patterning of the data and points the way to the problem's resolution. This feature of PBM will be compared with the role that problems play in empiricist inquiry in general and in the SUPER project in particular. To what extent did it include a problem focus, and what were the consequences of doing or not doing so?

On first reading, the SUPER project appeared to have a problem focus because its goal of the improvement of practice is suggestive of a discrepancy between an existing and a desired state of affairs. On further examination, however, it seems as if problems served little methodological purpose because inquiry started with the selection of research questions which were of interest to both participating researchers and practitioners, rather than with the identification and analysis of problems.

> While we were determining who at the university was interested in pursuing what questions, we were learning the high priority questions of the district-level administrators and school site participants. We stated at the outset that we would consider only those items that appeared on both lists, and that under no circumstances would we pursue any line that did not have committed participants at both the schools and University (Gifford & Gabelko, 1987b, pp. 390–391).

The implications of beginning research with a question rather than a problem can be seen by tracing the reasoning processes that may have led a researcher or practitioner to treat a particular question as significant. Take the following question which was on the list generated by the academics: "If students have an opportunity to do something useful for other people as part of their school work, will their motivation and performance in school improve, and will their future performance at work improve?" (Gifford & Gabelko, 1987a, p. 380). It seems reasonable to assume that the academics were interested in this question because they perceived a gap in current knowledge about student motivation and performance, and that the practitioners were interested in the question because they felt dissatisfied with the achievement of some of their students. Embedded in the question is a hypothesis that motivation and performance will improve if students are offered a different type of curriculum. This hypothesis, in turn, reflects a theory about student motivation, learning and effective preparation for the world of work. The empiricist emphasis on hypothesis testing is seen in the implicit transformation of an ill-structured problem (the improvement of motivation and performance) into a well-structured one (the problem is lack of work experience) whose solution qualities can then be tested via one or more hypotheses.

The swift transition from problem, to question, to hypothesis testing may be appropriate if other researchers have done the background descriptive investigation essential to theory development and there is a good match between the theory implied by the question and the local circumstances. If not, the solution expressed by the hypothesis may turn out not to be such, even if the hypothesis survives the testing process, because it does not satisfy the constraints of the original problem situation. In this case, the research process will have produced an answer which is not a solution to the practical problem. Haig (1987) makes the same point more generally by suggesting that talk of research problems in educational and social inquiry serves a largely rhetorical purpose.

> Methodological treatment of research problems that does exist typically amounts to the recommendation that we cast our research hypotheses in the form of questions.... [These demands] are frequently just requests for an operationalization of research hypotheses by way of an empirical specification of the relevant independent and dependent variables.... And, solutions to the original problems are thought to involve answering the questions by conducting experimental tests of the research hypotheses (p. 22).

In searching for questions of mutual interest, the researchers and practitioners overlooked the theories that generated those questions and their match to each other and to the problem situation. In addition, the collaboration represented by mutual interest in a question may be only skin deep, if practitioners and researchers have differing background assumptions about the significance of the questions. This brings us to the second point of comparison; the focus on theories of action.

Role of Theories of Action

The description of the research methods is sketchy, but it appears that reconstruction of theories of action, at least at the level of theory-in-use, was not a feature of the methodology employed in any of the reported projects. Achieving change in schools and practitioners was conceptualised as a matter of the adoption of the solutions provided by researchers, and the methodology did not make provision for the fact that those solutions might be incompatible with existing organisational or individual theories-in-use. In the following extract, Gifford and Gabelko (1987a) make a clear distinction between researchers, who provide solutions, and practitioners, who adopt and implement them.

> In order to be informed, effective practitioners, classroom teachers must have the opportunity to learn about valid and reliable solutions to classroom problems, as they are experienced by teachers, from those whose business it is to investigate and study classrooms and teachers—that is researchers (p. 368).

In Chapter 1 it was argued that such a theory of change is ineffective when researchers' solutions are incompatible with practitioners' existing theories. The methodology employed in SUPER did not incorporate procedures for examining and resolving any such incompatibility, because it was designed to evaluate the theory of the researcher and to bypass that of the practitioner. Despite the exclusion of practitioners' theories from the methodology, they intruded into the research in the form of challenges to the methodology itself. For the first eighteen months of the project there were numerous denunciations by practitioners of what they called the "psychostatistical paradigm" of the research. Gifford and Gabelko (1987a) explain what happened as follows:

> When the discussants were probed regarding their dislike/distrust of the psychostatistical paradigm, almost universally their opinions stemmed from negative experiences in educational psychology and other related "foundations" courses taken as part of their own teacher education program. It was in these courses that the future teachers and future administrators were first introduced to the tradition of research in education. And it was there that they determined that there was little relevance in it for them as classroom practitioners (p. 372).

The researchers responded to these challenges in the spirit of their collaborative commitment: they subjected themselves to lengthy questioning and responded to it with detailed justifications of their approach. The collaboration stopped short, however, of considering an alternative action research methodology which the authors believed would have been better received by practitioners. This dispute over method could not be collaboratively resolved because the researchers were unwilling to consider a methodology that required them to give up, as they saw it, control over the research process. Here is how Gifford and Gabelko (1987a) described the risks of action research:

> But there are risks and problems involved in action research: some easy to avoid through careful planning, and some beyond the reach of the authority of the Project and far into the realm of school district politics. Each of the risks and problems has the potential for halting the research. While many researchers are well aware of the potential benefits of conducting action research, few are willing to leave themselves so open to so much potential for "hostage taking", and remain wedded to other, less risky, models (p. 375).

The researchers were, therefore, in a dilemma; they wanted to be collaborative while at the same time retaining a methodology which severely restricted their ability to be responsive to the concerns of the practitioners. At the same time, a methodology which did not elucidate and appraise the theories of the practitioners allowed the latter to exercise considerably more authority over the researchers than may have been warranted. While the researchers guarded their methodology, practitioners, as we shall see in the next section, used their unexamined theories to make decisions about how and whether the findings delivered by the researchers would be implemented.

Theory of Change

The SUPER project incorporated an implicit theory of collaborative change that lies in stark contrast to that incorporated within PBM. It implies that change in schools and teachers can be brought about by delivering generalisations, tested within the context of application, to practitioners who had agreed to the importance of the research questions. The authors describe the limits of this type of collaboration as follows:

> If we were to build an operational definition of collaboration based on the experiences we have described, then collaboration is the process through which university researchers gain access, or permission, to work in tandem with school people. This is much better than not having access or permission. But it is not collaboration that can lead to institutional change. And the reason that collaboration as we know it is so limited (and limiting) brings us back to the research project director's question, "If SUPER were to disappear tomorrow, what would remain?" Read "disappear" to mean be without its own funding, and we have the real barrier to

> collaborative efforts that lead to institutional change (Gifford & Gabelko, 1987b, p. 419).

The limitations of collaboration are attributed to the politics of funding rather than to the limitations of their methodology and its associated social relations of inquiry and theory of change. Institutional change requires resources of time, money and expertise, but it also requires both individual and organisational learning. The methodology restricted practitioners' learning opportunities to learning about the generalisations produced by the researchers. When those generalisations did not cohere with practitioners' theories, the methodology could not adjudicate the differences, because it excluded the theories that resulted in the "resistance". The result for both practitioners and researchers was that learning was restricted to single-loop changes and that the dilemma between the acceptance and effectiveness of the research could not be resolved. These difficulties are intrinsic to the design of the research and cannot be overcome, as this project shows, by superimposing a collaborative research ethic onto a methodology that ignores practitioners' theories. The researchers' dilemma is captured in the following:

> We were secure in the proposition that mutually conducted research would have distinct advantages for practitioners. But the rest of the characteristics of "acceptable" new approaches were not that easy to meet:
>
> Compatibility with what practitioners know and do already,
> Degree of complexity—the lower the better,
> Trialability—that is, the ease with which practitioners try out the approach before accepting it,
> Observability—that is, the possibility to observe the approach working for someone else, as opposed to accepting it in the abstract before adopting it (Gifford & Gebelko, 1987a, p. 377).

To what extent are these criteria of research acceptability valued and attained within the PBM rubric? The first compatibility criterion expresses the acceptance–effectiveness dilemma which was discussed in Chapter 3. It is resolved in PBM by debating the implications for problem resolution of retaining or altering the source of the incompatibility. The problem focus of PBM provides a shared context in which decisions can be made about the degree of complexity which is required to solve a particular problem. There is no intrinsic value in complexity or lack of it; the level which is valued is that which is required by the particular problem. Similarly, some interventions will not meet the trialability requirement, because the problem requires a double-loop change, as in the cases reported in Part II. PBM offers a way of debating with practitioners how to reduce the risks of making such changes without sacrificing effectiveness. Finally, the observability criterion seems reasonable given the highly disruptive and demanding nature of many change efforts. While final results can neither be previewed nor promised, the improvability emphasis of PBM means that any alternative practices are made explicit, illustrated, modelled and subject to critique before and during the intervention phase of PBM. In short, many of the qualities of research which were valued by the teachers in the SUPER project are also valued in PBM and where they are not, the problem focus provides a context for debating the practical implications of their acceptance or rejection.

Summary

This chapter examined whether those features of PBM which made it well matched to problems and problem-solving were also present in empiricist methodologies, and whether their absence in any way jeopardised the practical contribution of empiricist research. Before proceeding with this comparison, a distinction was drawn between positivist and non-positivist forms of empiricism. Without this distinction, one runs the risk of stereotyping empiricist research and exaggerating the difference between it and PBM. For example, the verifiability principle, a central plank of positivist inquiry, would rule out inquiry into values and subjective experience, both of which are central to the analysis of many problems of practice. Non-positivist empiricism does not place such strictures on inquiry, however, since it does not accord a privileged place to observational evidence.

Despite the epistemological similarities between non-positivist empiricism and PBM, there are important procedural differences which can make them differentially effective in terms of their practical contribution. PBM puts considerable emphasis on the development of a holistic and accessible analysis of the problem. The emphasis on quantification and tight control in empiricist inquiry may lead to premature testing of a theory which fragments the problem or transforms it in ways that lead practitioners to reject its relevance. While a theory of an ill-structured problem is being developed, less expensive theory testing processes are needed which are more accessible to practitioners, give quicker feedback and encourage more cycles of theory development and theory testing.

Formal testing processes, as seen most clearly in experiments, incorporate social relations of inquiry which may be counterproductive in terms of resolving problems of practice. First, they focus on the theory of the researcher, not that of the practitioner, and it may be the mutual critique that each can provide that is essential to understanding and resolving a problem of practice. Second, the knowledge produced by experimental methods may not be applicable to the practice context, because, despite the attempt to force subjects to respond to experimenter-controlled influences, their reactions to those influences are the result of an unknown interaction between the theories of each party. The results of the experiment may have low predictive validity in situations where practitioners are free to respond differently.

The comparison process was continued via an example of empiricist inquiry which set out to improve aspects of teacher and school practice. It was argued that the lack of a problem focus, and the use of a methodology which bypassed the teachers' theories about both substantive and methodological issues, resulted in an impoverished form of collaboration and a limited practical impact.

References

Argyris, C., Putnam, R. & McLain Smith, D. (1985). *Action science: Concepts, methods and skills for research and intervention*. San Francisco: Jossey-Bass.

Bredo, E. & Feinberg, W. (Eds.). (1982). *Knowledge and values in social and educational research*. Philadelphia: Temple University Press.

Carr, W. & Kemmis, S. (1986). *Becoming critical: Education, knowledge and action research*. Lewes: The Falmer Press.

Editor (1987). Editor's introduction. *Education and Urban Society*, **19** (4), 363–367.

Evers, C. W. & Lakomski, G. (1991). *Knowing educational administration*. Oxford: Pergamon Press.

Gifford, B. R. (1986). The evolution of the school university partnership for educational renewal. *Education and Urban Society*, **19** (1), 77–106.

Gifford, B. R. & Gabelko, N. H. (1987a). Linking practice-sensitive researchers to research-sensitive practitioners. *Education and Urban Society*, **19** (4), 368–388.

Gifford, B. R. & Gabelko, N. H. (1987b). Research into research: SUPER project case studies. *Education and Urban Society*, **19** (4), 389–420.

Gitlin, A. D. (1990). Educative research, voice and school change. *Harvard Educational Review*, **60** (4), 443–466.

Habermas, J. (1984). *The theory of communicative action* (Vol. 1. *Reason and the rationalization of society*) (T. McCarthy, trans.) Boston: Beacon Press. (Original work published 1981.)

Hackman, J. R. (1985). Doing research that makes a difference. In E. E. Lawler, A. M. Mohrman, S. A. Mohrman, G. Ledford & T. G. Cummings (Eds.), *Doing research that is useful for theory and practice*, pp. 126–175. San Francisco: Jossey-Bass.

Haig, B. (1987). Scientific problems and the conduct of research. *Educational Philosophy and Theory*, **19** (2), 22–32.

Kaplan, A. (1964). *The conduct of inquiry: Methodology for behavioral science*. San Francisco: Chandler Publishing.

Lacey, A. R. (1986). *A dictionary of philosophy*. London: Routledge.

Lincoln, Y. S. & Guba, E. (1985). *Naturalistic inquiry*. Beverly Hills: Sage.

O'Hear, A. (1989). *Introduction to the philosophy of science*. Oxford: Clarendon Press.

Phillips, D. C. (1987). *Philosophy, science and social inquiry: Contemporary methodological controversies in social science and related applied fields of research*. Oxford: Pergamon Press.

Scriven, M. (1972). Objectivity and subjectivity in educational research. In L. G. Thomas (Ed.), *Philosophical redirection of social research* (The 71st Yearbook of the

National Society for the Study of Education, Part II), pp. 94–142. Chicago: University of Chicago Press.

Smith, J. K. & Heshusius, L. (1986). Closing down the conversation: The end of the qualitative/quantitative debate among educational inquirers. *Educational Researcher*, **15** (1), 4–12.

Walker, J. C. & Evers, C. W. (1986). Theory, politics and experiment in educational research methodology. *International Review of Education*, **32** (4), 373–387.

Walker, J. C. & Evers, C. W. (1988). The epistemological unity of educational research. In J. P. Keeves (Ed.), *Educational research methodology and measurement: An international handbook*, pp. 28–36. Oxford: Pergamon Press.

Weick, K. (1984). Small wins: Redefining the scale of social issues. *American Psychologist*, **39**, 40–50.

10

Interpretive Research and Problem Resolution

This chapter is concerned with a comparison between problem-based methodology (PBM) and the interpretive tradition of educational and social research. The question guiding the comparison is, "To what extent does the interpretive approach incorporate epistemological and methodological principles suited to the analysis of educational problems?"

To anticipate, I take the position that interpretivism incorporates features which are both well-matched and ill-matched to the purposes of problem understanding and resolution. The former are seen in the emphasis on the way actors construct their world through acting in accord with culturally shared linguistic and practical rules. This view of human action provides a rationale for investigating theories of action, and yields explanations of social life which employ the understandings of actors in the theories that researchers construct about them. The resulting accessibility and recognisability of such theories should be a major advantage over explanatory theories which bypass actors' understandings in promoting the self-reflection and education of such actors. Interpretive inquiry is ill-matched to practice, however, in that its educative and problem-solving possibilities are severely constrained by its limited criteria of theoretical adequacy, and by a functionalist stance which prevents the critique of actors' understandings. The interpretive tradition is of limited assistance, therefore, to those researchers who wish to move beyond explanation, to critique and problem resolution.

What is Interpretivism?

As for empiricism, generalisations about the major tenets of interpretivism are fraught by differences between and among those writers who make up the various strands of this tradition (Giddens, 1976, pp. 23–70). Interpretivists have in common, however, a view that the social and linguistic nature of their subject matter necessitates a form of explanation and inquiry that is qualitatively different from that employed in the natural sciences. A large part of the social sciences is concerned with the explanation of human action, and human action is, by definition, done with a point or purpose (Fay, 1975, ch. 4). An obvious implication of this idea is that one cannot correctly identify particular acts by reference to behaviour (or physical movement) alone. The observer must interpret the behaviour in terms of the attributed intentions, purposes or reasons of the actor in question. Fay (1975) uses the following examples to explain why behavioural evidence is an insufficient guide to the identification and explanation of action.

> For no physical movement is ever a necessary condition for an action—think of the myriads of ways in which one can vote, for example—simply because the aims which an action is intended to achieve can always be accomplished in literally countless ways; moreover, no physical movement can ever be a sufficient condition for a specific action to be said to have occurred because it is only in certain circumstances that particular movements can count as an action of a certain sort—thus, for example, saying "I do" in front of a priest and one's fiance, may be an act of marriage and it may not, depending on the circumstances, for the participants may be pretending or acting in a movie or rehearsing the ceremony, and so on. What specific action is being undertaken depends upon the meanings that the bodily movements being performed have (p. 72).

As the name implies, interpretivists explain action by interpreting it; that is, by retrieving the meanings embedded within it.

Given that reasons, beliefs and meanings are not directly observable in the manner of behaviour and physical objects, there is a problem in how these phenomena are to be reliably inferred. The answer given by most interpretivists is linked to the concept of *Verstehen*, a German term which can be roughly translated as "understanding" but which harbours a host of confusions and shifts in meaning. According to Giddens (1976, p. 19 and ch. 1), the concept of *Verstehen* associated with interpretivism up until about the 1960s (Dilthey and Weber being the major figures) involved a method of understanding social life through reliving or re-enacting the experiences of others. By getting "inside" the experiences of others, observers could empathise with them and thus gain access to the relevant mental states. Access to hidden mental processes could also be gained through sensitive encouragement of self-reports. The interpretations which such methods yielded were subject to the standard criticisms of subjective methods. People's reports of their intentions could suffer from lapses of memory, conscious and unconscious distortion and *post hoc* rationalisations. How could one distinguish between true and false self-reports and true and false attributions about others? More telling than these procedural criticisms, however, was the argument that this concept of *Verstehen* misconstrued the very notion of an intention. In Fay's (1975) words, notions like "intention", "meaning" and "motive"

> do not refer to occult processes hidden from the view of all but the individual person who is experiencing them and which cause the person's body to move in particular ways, but are rather ways of characterising the actions that we observe. Intentional explanations, for example, make sense of a person's actions by fitting them into a purposeful pattern which reveals how the act was warranted, given the actor, his social and physical situation and his beliefs and wants. An intention is no more "behind" the action than the meaning of the word is behind the letters of which it is composed . . . (p. 73).

These "purposeful patterns" are detected and understood because they are constituted by shared rules or conventions about how particular acts are accomplished. My intention to greet someone is accomplished and understood through its conformity to certain social practices and linguistic rules about greeting others. Thus, when someone in Western culture says "hello" it is understood as an act of greeting, not because listeners gain access to the mental processes behind the words, but because they recognise its conformity to a shared social practice. The observer has learned the significance of the behaviour through participating in the culture and thereby knowing the linguistic and contextual characteristics that make this an act of greeting.

In this second sense of *Verstehen*, largely associated with the publication of Gadamer's *Wahrheit und Methode* in 1960, the concept is removed from its subjective and individualistic connotations; understanding another is a matter of grasping the intersubjective meanings embodied in a language tradition. (See Giddens, 1976, pp. 54–64, for an account of this shift.) In contrast to the first sense, the evidence for understanding another is public rather than private. This is not to say that there are still not major difficulties in arriving at an accurate understanding, difficulties which will be dealt with subsequently. In the meantime, however, some discussion is needed of the implications of these notions of *Verstehen* for problem-based methodology.

Verstehen and Problem-based Methodology

The emphasis in PBM on explaining social and educational problems, in terms of the theories of action of relevant actors, appears quite similar to the type of explanation that is sought by interpretivists. In both cases, the aim is to show how actions follow sensibly from the particular understandings and purposes that the actor brought to the situation. Recall, however, that in PBM it is crucial to distinguish between two different types of theories of action because if they are incongruent for any given actor they will yield quite different understandings of that actor and of his or her role in a given problem situation. An espoused theory of action, because it is derived from an actor's self-reports, helps us to understand the self-understanding of actors. It may or may not help us understand the behaviours which their self-reports describe. A theory-in-use, the second type of theory of action, is derived from observing the behaviour of actors in a given type of situation and inferring the motives and understandings that must have informed it.

The distinction that is made in PBM between these two types of theory of action and the two types of understanding that they yield may be relevant to the previous discussion of the conceptual and methodological shifts associated with the concept of *Verstehen*. The earlier subjective concept of *Verstehen*, when accessed via self-reports, has its parallel in PBM in the concept of an espoused theory, for they both tell us how actors understand themselves. The later concept of *Verstehen*, based on the intersubjective meanings of particular language traditions, has its parallel in PBM in the theory-in-use. The importance of the distinction between these two types of understanding is easily overlooked when both are referred to by the somewhat ambiguous phrase "understanding an actor".

Both types of understanding are important within PBM, even though they may yield quite different explanations of behaviour. A theory-in-use captures the values and understandings that actually inform a person's behaviour; an espoused theory tells us how an actor actually understands their behaviour. The degree of congruence between them is a matter for empirical investigation in every case. Despite the possibility of incongruence, knowledge of an actor's self-understanding (espoused theory) is important in PBM because it represents the theory of the self which actors will use in their dialogue with researchers, and will thus reflect their view of the problem and of their own role within it. Knowledge of the actor's self-understanding helps to explain disagreements between actors and researchers; in addition, a marked discrepancy between self and others' understandings may, in itself, be part of the problem. If actors are implicated in a problem situation and misunderstand themselves, their self-reflections are unlikely to yield effective resolutions.

The Accuracy of Interpretive Understandings

Just prior to America's declaration of war on Iraq in January 1991, the following extract appeared in a *Newsweek* article entitled "Why We Can't Seem to Understand the Arabs".

> Consider the word hello. Iraqis who speak English use it when they mean goodbye. Or so it seems. "It was good to see you. Hello," they say. Or, perhaps, "You must leave. That's final. Hello." Foreign diplomats and journalists in Baghdad find this amusing. Perhaps, one suggests, it's not surprising Saddam hasn't gotten the message to pull out of Kuwait. We say goodbye, he says hello. Except—the Iraqis, in fact, are saying "ahlan wa-sahlan" in local dialect. It means welcome, whether you are coming or going. So American visitors blithely say goodbye, hear hello, and make jokes, while the Iraqis are saying welcome and wondering what the foreigners are sniggering about (Dickey, 1991, pp. 26–27).

The story illustrates some of the difficulties involved in attempting to understand others, particularly when observers and actors do not share a common linguistic and cultural tradition. The journalists and foreign diplomats understand the Iraqi use of Hello through their own Western understandings of the social practices that surround greetings and farewells. Their pre-understandings about such greetings and about how to react to such cultural and linguistic puzzles prevent them from making sense of this encounter.

One might take from this the lesson that interpretive researchers (and journalists and diplomats faced with interpretive problems) should set aside their own pre-understandings in order to make accurate inferences about unfamiliar social practices. But such a "tabula rasa" approach is not only psychologically impossible, it would preclude any sort of understanding, since we need prior concepts in order to even perceive and describe the actions in which we are interested. Interpretive understanding is inevitably the result of an interaction between our pre-understandings and the understandings and practices which are the subject of our inquiry. As Denis Phillips (1991) puts it:

> ... interpreters who are attempting to grasp the meaning of an actor, or to grasp meaning that has been objectified in some way, have *their own* understandings shaped by the fact that they themselves are members of a particular culture at a particular historical moment. Interpretation, in other words, is not an act in which a "disembodied" investigator is trying to decipher the (pre-established) meaning of a culturally and historically situated actor

> or institution; rather, the interpreter, too, must become hermeneutically aware of his or her own historicity or "preunderstanding" as some writers term it (pp. 555–556).

If interpreters bring different pre-understandings to their inquiries, then they will produce different if not conflicting interpretations. How are we to adjudicate between them? Two criteria are mentioned within the interpretivist literature. The first is a notion of coherence; that an adequate explanation makes clear how certain actions make sense given the particular meanings that actors hold about themselves and the situation. Taylor (1977) writes that we make sense of an action "when there is coherence between the actions of the agent and the meaning of his situation for him" (p. 109). So the correct, or more correct, interpretation of the Iraqis use of "Hello" is that which eliminates the puzzle by showing how the Iraqi greeting is used to both welcome and farewell.

A second criterion concerns the response of the subjects to the interpretivists' account. An adequate account provides the knowledge needed to sustain encounters with those it purports to understand, whether or not that capacity is actually employed (Giddens, 1976, p. 149). This is judged by the response of the lay members of the community "in so far as they are prepared to accept what the observer does or says as 'authentic' or 'typical'" (p. 149). Giddens acknowledges, however, that disagreement is not an automatic indicator of inadequacy.

> . . . there may be characterizations of an actor's conduct that he may not only find unfamiliar, but which he might actively refuse to recognize as valid if presented with them. The latter circumstance is certainly not a sufficient basis in and of itself to reject them, although how far he "understands" them, or can be helped to understand them, and how far he accepts them as characterizing his own conduct is very likely to be *relevant* to adjudging their accuracy (p. 150).

The limitation of both these criteria is that what counts as making sense or as coherent, is itself determined by the framework of pre-understandings that each party brings to the situation. Both the interpretive explanation and the judgement of the sense of that explanation are interdependent. It is this "hermeneutical circle" which will prevent the resolution of many disagreements about the adequacy of interpretive explanations.

> Making sense in this way through coherence of meaning and action, the meanings of action and situation cannot but move in a hermeneutical circle. Our conviction that the account makes sense

> is contingent on our reading of action and situation. But these readings cannot be explained or justified except by reference to other such readings, and their relation to the whole. If an interlocutor does not understand this kind of reading, or will not accept it as valid, there is nowhere else the argument can go (Taylor, 1977, p. 109).

If it is true that there is nowhere else to go beyond our context-bound view of coherence and our context-bound attempts to convince others of the adequacy of our views, then there are substantial difficulties in store for PBM. The type of dialogue, critique and learning that is central to the methodology requires that we can see, under particular conditions, beyond our conceptual frameworks. I shall argue that Taylor's conclusion that there is nowhere else to go is unduly pessimistic, because it exaggerates the closedness of the hermeneutic circle and neglects the role of empirical criteria in theory adjudication.

I should acknowledge at the outset that there is much merit in the idea of the hermeneutic circle; the theory-ladenness of observations and the role of prior conceptualisations has been recognised in the natural sciences at least for the last twenty years (Phillips, 1990). Similarly, the necessity of some circularity in the analysis of ill-structured problems was discussed in the second chapter of this book. My disagreement comes with the degree to which inquirers are portrayed as unable to break out of these pre-understandings so that the circularity becomes vicious. One way to avoid such viciousness is to treat the competing interpretations as hypotheses and to construct mutually acceptable tests of the points of difference. Returning to the Iraqi example, imagine that the journalists in question did not accept the interpretation that "Hello" was being used as a generic greeting, rather than as a welcome. They might argue that the Iraqis were using the term in a deliberately confusing manner, because they hated foreigners, especially Americans, and they were trying to make their stay in Iraq as uncomfortable as possible.

If the latter explanations were the correct one, then one would predict that Iraqis who did not dislike Americans would not use Hello in the way described. If it turned out that they did, then the explanation based on an attempt to confuse foreigners would be discredited. Now those interpretivists who agree with Taylor's views might argue that the opponents would never agree on a "mutually acceptable" test, and that even if they did, their differing pre-understandings would lead them to interpret the results differently. My reply would be that the existence of some differing pre-understandings does not entail that all pre-understandings are different. There are, after all, probably dozens of ways in which the two accounts are similar; they both locate the difficulties in the interaction between two different cultural groups; they identify the groups in the same way, etc. Resolution through examining relevant evidence can be achieved if some of the shared pre-understandings can serve as common ground against which to examine the implications of the differences. If such a touchstone cannot be found, because one or more of the parties insists that all possible candidates are unacceptable, then it seems reasonable to apply the criterion of improvability, discussed in Chapter 2, and to give less weight to interpretations that use their hermeneutical qualities in ways that make them self-sealing and untestable.

In conclusion then, the hermeneutic circle should not be conceived as a cage, preventing either articulation with other circles or its own revision. Interpretations do depend on prior understandings, but they are not determined by them; nor are they so idiosyncratic to particular cultures that points of overlap cannot be found which serve as empirical common ground between them, against which the implications of particular differences can be traced and resolved. PBM incorporates interpretivist principles about the importance of actors' meanings and the interactive processes involved in discovering them, while rejecting those accounts of the hermeneutic circle that would make it impossible to rationally resolve disagreements about the accuracy of researchers' accounts.

Interpretivism and Going Beyond the Status Quo

In a paper entitled "Can Ethnographic Research Go Beyond the Status Quo", Courtney Cazden (1983) tells a story about a sign in the Alaskan State Department of Education that read:

WE DON'T NEED ANY MORE
ANTHROPOLOGICAL EXPLANATIONS
OF SCHOOL FAILURE

The explanations to which these educators were referring are those provided by ethnographies, a type of interpretive inquiry that has its origins in anthropology (Taft, 1988; Wolcott, 1988). Their plea, according to Cazden, was for research that went beyond description and explanation to intervention. In this section, I will examine the extent to which the interpretive tradition provides educational researchers with the theoretical and methodological tools to do what the Alaskan educators were calling for. I shall argue that the tradition does incorporate an implicit theory of change, but that this theory is inadequate to the task of intervention in most types of educational problem.

The distinction between educational change and problem resolution, discussed at length in Chapter 1, is relevant to our evaluation of the intervention possibilities of interpretive research. Interpretivism has the potential to promote change by increasing the possibility of communication between groups who previously misunderstood one another, or who were dismissed as strange or irrational (Fay, 1975). Once we know the meanings of their social and linguistic practices, it is possible to engage in dialogue, and out of such dialogue may come a variety of practically significant alterations to inter-group relations. Interpretivism also holds the potential for greater understanding of self, either through comparison with the practices of others, or through debate about discrepancies between one's self-understanding and the analysis presented by the researcher. Provided that the subjects of the analysis judge it to be accurate, the latter form of dialogue has the potential to increase actors' understandings of the rules and conventions that pattern their behaviour, and to reflect more accurately about their desirability. As Cazden (1983) argues, however, this type of change process is only likely to make an impact on the simplest of educational problems. She asks, ". . . what is the convincing argument or evidence for what seems to me our almost exclusive reliance on raising the consciousness of practitioners as the process by which our description of the status quo can lead to change?" (p. 34). The reasons for this limitation are the restricted scope of interpretive explanations and the lack of a normative standard against which to judge the social practices under investigation.

Interpretive explanations champion the role of human reasoning, motives and purposes in the production and reproduction of social practices. For example, an interpretive researcher might describe the way "at risk" children are identified, processed into special programmes and submitted to a slow-paced curriculum, because their teachers believe that they cannot keep up with the regular curriculum, and that they arrive at school with significant deficits. Such meanings as "cannot keep up" and "significant deficits" are important components of an explanation of the observed teaching practices, but they need to be supplemented for the purposes of problem resolution (and adequate social theory) with an explanation of their origins and persistence. To what extent do policies about funding, textbooks and assessment practices sustain and encourage such beliefs? Explanations that do not ask about the links between actors' beliefs and relevant institutional practices are liable to the criticism of idealism. In this technical sense, a theory is idealistic when it implies that people's ideas are the sole cause of particular practices, and that all that is needed to alleviate social problems is alteration of those ideas. Fay (1975) describes this limitation as follows:

> . . . such a social science leaves no room for an examination of the conditions which give rise to the actions, rules and beliefs which it seeks to explicate, and, more particularly, it does not provide a means whereby one can study the relationship between the structural elements of a social order and the possible forms of behaviour and beliefs which such elements engender (pp. 83–84).

In order to avoid the same criticism, PBM also needs to incorporate social conditions as well as actors' meanings into its explanatory apparatus. If it does not identify the conditions which sustain problematic beliefs and practices, its problem analyses will yield ineffective action strategies, where those beliefs cannot be altered independently of relevant social conditions.

The second, and probably more important, cause of the limited contribution of interpretive research to educational problem resolution is the absence of a normative standard against which researchers can judge the adequacy of the understandings and social practices which they seek to explicate. Two such standards are central to PBM: the degree to which actors' understandings and practices contribute to the focus problem, and the degree to which those same features facilitate or enhance learning and problem-solving. These standards are the basis for critique and for the design of alternative understandings and practices. While interpretive researchers are concerned about the accuracy of their portrayals of social practices, they are not interested in judging those practices; in fact, such judgements might be viewed as inimical to the sympathetic and authentic portrayal of their subject matter (Wolcott, 1988, p. 203). By showing how social practices make sense, given the understandings of those involved, interpretivists portray social life as adaptive and functional. Each practice is functional when looked at in its own context, and the result is an explanation which portrays social practices as tightly interlocked, and actors as making choices which seem inevitable.

The problem for the practitioner is that many educational decisions require choices to be made between apparently conflicting but internally coherent understandings. The interpretive approach helps such practitioners to understand why the parties hold the views they do, but provides no help in making a rational choice between them (Evers & Lakomski, 1991, p. 10). If mutual understanding and consciousness-raising prove insufficient, interpretive research can provide little else. Questions about the validity of competing beliefs, and about the intended and unintended consequences of enacting those beliefs, are not asked by the interpretivist. It is the answers to these questions that provide the basis for critique, for making rational choices between social practices and for redesigning those that are problematic. Without these tools, interpretive researchers will be forever limited to describing the status quo, and to fostering change through highly unobtrusive, non-evaluative strategies.

An Example of Interpretive Research

The work of Douglas Campbell and his colleagues at Michigan State University (Campbell, 1988) provides an excellent context for further discussion of the practical contribution of interpretive methodology, because the author identifies strongly with this tradition, and the research is concerned, as in the previous empiricist example, with the improvement of practice through the development of collaborative relationships with teachers. In addition, the model of collaboration which the research incorporates has been criticised in a subsequent publication (Ladwig, 1991), and the implications of that criticism for PBM should be examined.

Campbell's paper reflects on a three-and-a-half-year collaboration between teachers and researchers designed to "... implement and study an approach to staff development that engages participants in reflection and dialogue grounded in respect for teachers' knowledge and modeled on the nonjudgmental and nonintervening features of ethnographic inquiry" (p. 99). The particular focus of the staff development efforts was chosen by teachers themselves, and once selected, researchers used a range of ethnographic techniques to both document the process and to stimulate teachers' critical reflection on their own practice.

> ... the researchers assumed the role of participant-observers in the teachers' classrooms and in meetings of all project participants. For the purpose of documenting project activities, the researchers collected observational field notes, made audio and video recordings, and conducted informal interviews with teachers. These materials have also been essential to the researchers' support of the teachers' efforts to reflect on their practice, through their use as a basis for dialogue in journal exchanges and for group discussions among all participants about project goals, directions and substantive educational issues (p. 102).

Campbell describes this approach to staff development as one of engaging teachers in an ethnographic research process, and contrasts it with the more traditional approach to staff development involving the dissemination and translation of research findings. He believes this latter approach is frequently ineffective for many of the same reasons as were discussed in Chapter 1.

Problem Focus

To what extent was the project concerned with understanding and attempting to resolve an educational problem? At one level, the problem was how to improve the professional development of teachers. At another more specific level, the problem was the topic, such as reading instruction, which each teacher chose as the focus of his or her professional development activity. To have a problem focus, as defined in PBM, however, involves far more than the identification of a problem. It involves, in addition, the analysis of the problem by developing a theory which explains how it arises and points the way to its resolution.

We cannot judge the extent to which Campbell's team attempted to analyse each teacher's problem prior to intervening, because so little description is provided in his report. An analysis of the wider problem of professional development is proposed, however, in terms of Campbell's belief that teachers "typically do not have sufficient time, resources, and reward to do the deep reflecting required to draw on [their] knowledge in identifying and implementing changes appropriate for what they know best about their children's learning needs" (p. 102). This analysis suggests that the problem can be improved by the provision of time, resources and reward, so that teachers can reflect about and improve their instructional practices.

This problem analysis differs from the type provided in PBM because it presents no evidence about the capacity of the system itself to provide the conditions which Campbell suggests are essential to resolving the problem. This capacity should be evaluated in a problem analysis, because if it is not present, the intervention will only be effective while researchers are actively involved. In summary, while Campbell's research does focus on an educational problem, the analysis differs from that which would be conducted in PBM, because it remains disconnected from the beliefs and practices of those who are responsible for the conditions which are alleged to be causally involved in the problem.

Role of Theories of Action

One would expect research conducted within the interpretive and ethnographic traditions to concentrate on uncovering the beliefs and values that underpin patterns of teacher practice. This was certainly true, in general terms, in Campbell's study, because helping teachers to gain access to and critically reflect upon their knowledge of practice was the means to improvement. The process differed from that of PBM, however, in that the researchers did not explicitly evaluate the adequacy of teachers' theories of practice. One reason why they did not do so, was that Campbell believed that teachers had the capacity to improve their own practice with minimal outside intervention. The view that people have this capacity for self-improvement is also seen in some types of counselling theory (Rogers, 1961) and in our common-sense assumptions about people's capacity to learn from their experience. Equally, the contrary view is supported by some empirical and theoretical work on change in interpersonal and organisational contexts. Argyris, for example, would argue that our self-managed change efforts are only successful for single-loop problems; namely, those that can be resolved without challenging the assumptive framework within which we think about the problem in question (Argyris, 1983). The resolution of double-loop problems requires an outsider, not just to facilitate, but to critique the assumptive framework being brought to the problem-solving process. It is impossible to resolve the issue in the context of this particular study, because Campbell does not present any evidence about the impact of his intervention on teachers' learning.

A second related reason why the researchers avoided a critical role was that it would violate their espousal of the "non-judgmental and non-intervening features of ethnographic inquiry" (p. 99). If teacher knowledge was to be critiqued, it would be done by the teachers themselves, stimulated by the resources which the researchers provided or by their open-ended, facilitative inquiries. More will be said about this in the next section on the theory of change.

Theory of Change

The theory of change that is revealed in Campbell's approach to teacher development provides a clear contrast to that of PBM. Change in PBM involves uncovering the theories of action of practitioners involved in the problem situation, evaluating their implications for resolving it and, if necessary, learning a different theory of practice. While all of these tasks may have been accomplished under Campbell's model of staff development, it was not designed to ensure that they were, because this would have required the researchers to take a far more active and evaluative role than is compatible with the principles of ethnographic inquiry. The role of the researcher was to stimulate and deepen the teacher's reflections by:

> . . . modelling a style of open-ended and nonprescriptive questioning and reflection drawn from the traditions of anthropological fieldwork and adapted to the goal of empowering teachers to look more deeply at their own practice and to make their own critical judgements and decisions about whether, what, when, and how to change their teaching (p. 109).

In evaluating this theory of change, we are simultaneously concerned with the explanatory accuracy of his theory of the problem (that it is caused by inadequate resources for reflection) and the effectiveness of his theory of intervention (the provision of resources to promote teacher reflection through the non-judgemental processes of ethnographic inquiry). The only reference to project outcomes is an unsubstantiated claim that "the project has largely succeeded in fostering significant changes in the teachers' and principals' thinking, practices and self-esteem ..." (p. 99). Assuming that some changes were made, it does not follow that problems were solved, or that the intervention was responsible. In fact, there is some doubt about the latter, because much of Campbell's article is devoted to a discussion of the difficulties in implementing the theory of intervention which the researchers espoused. I shall argue that this theory could not be implemented because it incorporates an incoherent role for the researcher and an incoherent concept of collaboration.

Campbell spends much of the article documenting the way in which the researchers failed to act consistently with the non-evaluative and non-judgemental role which they espoused. For example, when some teachers nominated classroom management as their professional development topics, the researchers attempted to reframe them as issues of curriculum and pedagogy. Eventually they were confronted by the teachers for being unable to be upfront about their preferred professional development agenda. The researchers violated their own non-evaluative precepts in two ways. First, as Campbell acknowledges in his article, it is impossible to ask neutral questions, and to facilitate reflection, without a theory of what it is important to ask about, and what might count as worthwhile reflection. Second, and Campbell seems unaware of this point, there is a contradiction between the proscription on evaluation and the researchers' judgements that the "practitioner's viewpoint has validity in its own right, and should prevail in decisions about staff-development directions" (p. 105). This highly evaluative judgement suggests that the proscription is on critical judgements, not on judgements in general.

Despite Campbell's recognition that the neutral, non-evaluative ideal of ethnographic inquiry is unattainable and even contradictory, he renews his commitment to it, and reports that the team subsequently tried to be more attentive and sensitive in its future practice (p. 112). The reasons for this puzzling choice are related to his concept of collaboration and its relationship to evaluation. Campbell's concept of collaboration is not entirely clear, because it is variously discussed in terms of the initiation of tasks (p. 105), the division of labour between researchers and practitioners (p. 104) and the control over decisions. It is the latter concept which seems to predominate, but even given a decision-based concept of collaboration, Campbell seems unsure as to whether it requires practitioner control or a shared responsibility for project decisions. For example, the previous quotes about the validity of practitioner knowledge and their control over decisions about whether, when and how to change their teaching, suggest that Campbell envisages collaboration as practitioner control over key project decisions. This concept of collaboration is inadequate, however, because it implies unilateral control of researchers by practitioners, rather than bilateral control over key project decisions.

It is now easier to understand why Campbell wants to hang on to his ethnographic ideals in spite of his acknowledgement of their paradoxes and contradictions. He rightly sees the traditional model of teacher development involving the delivery of research findings to teachers as disempowering and ineffective. He mistakenly concludes from this, however, that the only choices are between the delivery of answers via research findings and the non-judgemental and non-directive ideals of ethnographic research. PBM provides an additional option which also engages teachers in a process of inquiry, but without requiring the abandonment of critical judgement and the presumption of the validity of teacher knowledge. What PBM does require is precisely what Campbell's teachers were calling for, namely that researchers be upfront about their judgements; that they place themselves in the process of inquiry rather than stand outside it, so that their own knowledge can contribute directly and openly to mutual evaluation and improvement. By adopting this approach to change and teacher development, Campbell could preserve those features of ethnographic inquiry that contribute to the improvement of practice, such as its emphasis on teachers' theories and on ecological validity, while abandoning those that are counterproductive to it, such as its relativist epistemology which precludes critical dialogue between teacher and researcher.

Ladwig's Concept of Collaboration

Ladwig (1991) is also critical of Campbell's concept of collaboration, but from a very different perspective from that of PBM. His rival theory of collaboration provides an opportunity to further examine the adequacy of the researcher–practitioner relationship that is advocated in PBM.

The position Ladwig takes is that the researcher–practitioner relationship that Campbell describes is exploitative rather than collaborative. He reaches this conclusion by first presenting a model of exploitation, based on the work of John Roemer and Erik Olin Wright, and then arguing that it fits Campbell's example. One group is said to be exploited by another when it experiences a situation of oppression, which the oppressor benefits from by the transfer of value generated by the oppressed group. The oppressed group is in a situation of oppression when they are worse off than in a feasible alternative, which the oppressor prevents them from attaining.

Ladwig then attempts to show, using Campbell's example, that these are the conditions under which so-called collaborative research in education is produced. The labour of the teacher is essential for the conduct of such research, but the value which is generated is disproportionately accrued by the researcher. For example, of the four types of capital that may be gained from such research (associated with career advancement, authorship, methodological skills and improved instructional knowledge), only the latter is generally available to teachers. In the feasible alternative, conditions of employment, including career structures, would be altered so that teachers could gain capital value from all four benefits. Campbell then argues that researchers prevent teachers from entering this alternative by failing to systematically advocate for it. The blurring of the roles of teacher and researcher that would result would be to the latter's disadvantage because it would increase the competition for the benefits that accrue from research. He concludes that asking teachers to engage in research, without enabling them to equally profit, or striving to make this possible, is exploitative.

Ladwig's argument hinges on the reasonableness of equating exploitation with preventing others (in a very weak sense of prevent) from deriving equal material benefits from shared research activity. His analysis assumes that material benefits are the only type of benefit relevant to the issue of exploitation, and that the decision process under which the relationship is entered is irrelevant to the judgement of exploitation. On the first point, non-material benefits may be equally or more important to the participants than material ones, and if Ladwig is going to insist that only the latter are relevant, then he has to also argue that the subjective assessment of benefit is irrelevant to the issue of exploitation. For example, in contracting with Tony and Carol, both indicated that their motivation to participate in the research turned on the opportunity to learn new skills and to reflect about their jobs with the researchers. Being university graduates, they were aware of the possible career benefits that the research might bring to the author, but this did not lead them to want to participate in those benefits, or to see themselves as exploited because they could not gain similar career opportunities from their participation. Carol and Tony made their decision by evaluating the likelihood that the project would deliver the benefits, whether material or non-material, that were important to them. The question of whether or not the relationship was exploitative turns not on the relative distribution of a prespecified set of benefits, but on the conditions under which the decision to participate was made, and on the sincerity of the researchers' efforts to deliver whatever benefits were mutually agreed. This brings us to the second critical point: namely Ladwig's ignoring of the decision process.

A decision to enter and sustain a research relationship cannot be exploitative, even if unequal benefits accrue, if parties make it based on all available relevant information, and if they are free to leave the relationship if circumstances change. A frequent objection to this consensual, decision-oriented concept of non-exploitation is that free choice is impossible when there are power and status inequalities between the parties. But what is the relevant concept of power behind this objection? It cannot be that differential capacity to exercise power precludes consensual decisions, and thereby collaboration, because this concept would practically rule out the possibility of free uncoerced choice. The threat to genuine collaboration comes when decisions are distorted by the actual exercise of power, and so a decision-based analysis of collaboration must incorporate an analysis of the prevention and detection of such distorting influences. Such an analysis was incorporated in the prior discussion of critical and unilateral dialogue (Chapter 3).

In summary, both Ladwig (1991) and I have grave doubts about the model of collaboration which Campbell proposes. Ladwig's objections are mistaken, however, because collaboration and non-exploitation are concerned with notions of consent and autonomous decision-making, not with relative material benefit. My criticism of Campbell's concept of collaboration is that it attempts to redress power imbalances in research relationships by substituting practitioner control over research for the more typical, equally problematic, reverse pattern. This results, as Campbell himself acknowledges, in contradictory, inefficient and even manipulative relationships, which jeopardise the possibility of reaching a shared understanding of educational problems and of how to resolve them.

Summary

Problem-based methodology and interpretivism both give a central place to the way actors make sense of a situation and the way those meanings are reflected in their actions. In PBM, however, a careful distinction is made between understanding an actor and understanding the self-understanding of an actor. The former is derived from studying what people do, the latter from the actor's self-reports. This distinction has often been blurred in discussions of *Verstehen*, and a false competion set up between them. In PBM, both types of understanding are considered essential, since if the two are incongruent, the former will have far more predictive power than the latter. On the other hand, actors will bring their self-understandings into any dialogue, so a collaborative approach to problem understanding and resolution cannot be achieved without acknowledging and resolving any such discrepancy.

Interpretive inquiry has been criticised for its sole reliance on an internal explanatory apparatus. The conditions that generate and sustain actors' understandings are given little attention, at least in the theory of interpretivism. In PBM, the theory of action is linked to the external conditions which sustain it, through conceptualisation of organisational processes (for example), as organisational theories of action which are the context for individual action. The inclusion of such factors is essential to an explanatory analysis and a programme of problem resolution that avoids the charge of idealism.

The relativist epistemology of interpretivism severely limits its critical and intervention possibilities. Critique must proceed via an explicit standard, and such evaluative inquiry is counter to the non-judgemental stance of most interpretivists. Intervention builds on critique by showing how alternative theories and associated practices may lead to more desirable social arrangements. Since interpretivist methodology gives little emphasis to such normative theorising, its intervention possibilities are limited to those associated with the enhancement of mutual and self-understanding, and these are seldom sufficient for the resolution of long-standing educational problems.

As we saw in the illustrative example, interpretivist researchers who do wish to make a difference may be caught between their non-evaluative espousals and their attempts to shape practice according to their undisclosed normative theories. The result may be at best an inefficient wastage of researcher expertise, or, at worst, manipulation of practitioners in the direction of implicit and unexamined critical and normative standards. Suppression of the researchers' theoretical and critical expertise, whether in the interest of non-evaluative ethnography or of practitioner development, jeopardises the possibility of the critical engagement required for problem resolution.

References

Argyris, C. (1983). *Reasoning, learning and action.* San Francisco: Jossey-Bass.

Campbell, D. (1988). Collaboration and contradiction in a research and staff development project. *Teachers College Record*, **70** (1), 99–121.

Cazden, C. (1983). Can ethnographic research go beyond the status quo? *Anthropology and Education Quarterly*, **14** (1), 33–41.

Dickey, C. (1991). Why we can't seem to understand the Arabs. *Newsweek*, January 7, pp. 26–27.

Evers, C. & Lakomski, G. (1991). *Knowing educational administration.* Oxford: Pergamon Press.

Fay, B. (1975). *Social theory and political practice.* London: Allen & Unwin.

Giddens, A. (1976). *New rules of sociological method.* New York: Basic Books.

Ladwig, J. G. (1991). Is collaborative research exploitative? *Educational Theory*, **41** (2), 111–120.

Phillips, D. C. (1990). Post positivistic science: Myths and realities. In E. G. Guba (Ed.), *The paradigm dialogue*, pp. 31–45. Newbury Park, CA: Sage.

Phillips, D. C. (1991). Hermeneutics: A threat to scientific social science? *International Journal of Educational Research*, **15** (6), 553–568.

Rogers, C. (1961). *On becoming a person: A therapist's view of psychotherapy.* Boston: Houghton Mifflin.

Taft, R. (1988). Ethnographic research methods. In J. P. Keeves (Ed.), *Educational research methodology and measurement: An international handbook*, pp. 59–63. Oxford: Pergamon Press.

Taylor, C. (1977). Interpretation and the sciences of man. In F. Dallmayr & T. McCarthy (Eds.), *Understanding and social inquiry*, pp. 101–131. Notre Dame, Indiana: University of Notre Dame Press.

Wolcott, H. (1988). Ethnographic research in education. In R. M. Jaeger (Ed.), *Complementary methods for research in education*, pp. 187–206. Washington: American Educational Research Association.

11

Critical Research and Problem Resolution

Critical research is attractive to many educational researchers because they perceive it as overcoming the limitations of both the empiricist and interpretive traditions. Critical researchers reject the political and value neutrality of positivist versions of empirical inquiry, arguing that such a stance only serves to hide the values and interests which are, perhaps unwittingly, served by research which makes such claims. The question for them is not *whether* research is value free, but *which* values and politics are promoted by it. The goal of critical research and the variety of critical theories which inform it is to promote a particular politics; namely, one devoted, as Braybrooke (1987) describes it, to "emancipation—emancipation of social classes, from oppression or contempt; emancipation of people throughout society, from ideas that inhibit rationality" (p. 68).

Critical theorists also see their approach as overcoming the idealism and relativism of interpretivism. The former is overcome by combining explanations which draw on subjective understandings with causal accounts of the external conditions which give rise to and sustain them. Such explanations enable critical researchers to critique such understandings by showing the reflexive unacceptability of both their origins and their underlying interests (Geuss, 1981, ch. 3). The relativism of much interpretive inquiry is overcome in critical research by incorporating an explicit normative theory, which sets out the principles through which one can critique current social arrangements, whether at the societal, organisational or interpersonal level.

Critical theory poses a considerable challenge to PBM because both approaches share the same practical purposes and the same broad procedural framework for achieving them. For example, both begin with a problem situation, investigate the understandings of relevant actors, critique those understandings and show how alternative understandings and actions can resolve problems consistently with a normative standard. Despite these surface similarities, I shall argue that there are important differences between PBM and critical theory, and that these differences should make PBM more effective than critical theory in achieving their shared practical purposes. In brief, my argument will be that much critical research in education falls short of its practical promise because the scope of the problems it tackles is too large, its critical analyses fail to show how particular agents in particular situations can act to resolve their situation, its educative strategies frequently omit the powerful, and it has given too little emphasis to the development of an effective theory and practice of interpersonal and organisational change. On the other hand, a systematic comparison of PBM's theory of critical dialogue with Habermas's theory of rational consensus would be fruitful for a programme of normatively-based empirical research on the micro-politics of educational problems and their resolution.

Considerable groundwork must be laid before these arguments can be fully explicated and defended. This chapter will proceed, therefore, with a brief introduction to critical theory and to its methodology, before returning to the nature and implications of some of its differences with PBM.

What is Critical Research?

Some may consider it premature to describe critical research as a tradition in educational research, because the history of the method, in education at least, is little more than thirty or forty years old (Young, 1989, p. 62). During that time, however, it has gained some of the accoutrements of a tradition, with its regular inclusion in theoretically inclined treatments of social science and educational methodology (Braybrooke, 1987; Bredo & Feinberg, 1982; Carr & Kemmis, 1986), in the appearance of reviews of critical research (Anderson, 1989; Ewert, 1991) and in the emergence of a series of critiques of the method itself (Lakomski, 1988; Walker, 1985).

Offering a precise definition of the method is extremely difficult, because the terms critical research and critical theory refer to both a methodology and to a loose grouping of social theories. Given that this book is primarily about methodology and not about social theory, I could confine myself in this chapter to critical research in its methodological sense. Such a separation is not entirely justified, however, because the methodology is inextricably linked with a theory of persons and of the societal, institutional and interpersonal conditions needed for their fulfilment.

As methodology, critical research is dedicated to the understanding and alteration of those conditions which prevent people from living fulfilling and satisfying lives. The starting point of a critical project is the frustration or unhappiness of a group of people and an analysis of their suffering in terms of the conditions, including their self-understandings, which maintain their unhappiness. This analysis phase is followed by a period of education (ideology critique) in which false understandings are replaced by alternatives which better serve the real interests of those who are the subjects of the research. Finally, a programme of social action, designed to transform the original situation, is pursued and the results evaluated (Fay, 1987, ch. 2; Geuss, 1981, ch. 3; Young, 1989, p. 41).

As described, the explanatory–diagnostic purposes of critical research can be informed by any social theory which offers an account of why particular conditions of oppression and irrationality have arisen and are sustained. Writers like Brian Fay (1987, pp. 4–7), for example, see the theories of Laing, Habermas and Marcuse, as well as those of Marx, as potentially appropriate to this task. Other writers want to argue that critical inquiry must be informed by a much narrower range of social theories, namely those associated with a neo-Marxist critique of advanced capitalist economy. Comstock (1982, p. 382), for example, makes such a theory a defining characteristic of critical inquiry.

Sorting out the variety and complexity of issues involved in deciding what counts as an appropriate critical theory is well beyond the scope of this volume. However, for the purposes of structuring the comparison with PBM, two major groupings of critical theory will be distinguished. The first group comprises those theorists who appeal primarily to Marxist concepts to explain conditions of disadvantage, alienation and social conflict. For writers like Comstock (1982), and social reproductionist and correspondence theorists of schooling (Bowles & Gintis, 1970), the key to understanding these conditions lies in the technology and social relations of the capitalist mode of production and its associated ideologies and class conflict. Schools are understood as sites for class dominance and struggle, not for the free development and fulfilment of human beings. If educational problems are largely attributable to capitalist social relations, then the success of educational interventions will be constantly constrained by the wider economic context.

The second major grouping of critical theorists are those associated with the Frankfurt school, an interdisciplinary group of social theorists whose major contemporary exponent is Jurgan Habermas. The group engaged in a thoroughgoing revision of Marx's political economy, intended to account for the fact that the proletarian revolution which Marx had predicted had failed to eventuate, despite favourable conditions. Its various members argued that the tendency towards fascism in Europe, the intrusion of the state into increasing areas of public life and the involvement of the working class in reactionary rather than revolutionary movements required an explanatory apparatus that was broader than that provided by Marx's political economy (Thompson & Held, 1982, pp. 2–3). The cultural and social, as well as the economic, had to be brought into the framework of critical theory so that one could explain how it was that social fragmentation and alienation persisted despite relative material affluence.

Part of Habermas's revision of Marx has been to argue that we make our history through communication and culture as well as through our work. A major cultural and communicative force which has thwarted human progress is the inappropriate application of an instrumental form of reasoning, which has allowed humankind to so successfully dominate nature, to areas of human and social activity. This narrowly conceived form of rationality is concerned with administrative procedures, efficiency and taken-for-granted ends rather than with the desirability of the ends and the standards used to judge the means. The acceptance of this form of reasoning as the only form of rationality relegates the search for consensually defined forms of authority, standards and shared meanings to the realm of arbitrary and relativist decision arenas. The result is an impoverished public domain, loss of community and the alienation of citizens from the institutions which control their lives. Habermas's thirty-year project has been to reveal the one-sided nature of this form of rationality and to argue for a much broader concept, which can be the basis of emancipatory critique.

This brief and highly simplified introduction to Marxist and Habermasian critical theorists suggests that the two major groupings differ in the incorporation by the latter of non-materialist explanations of social problems and in their theories of social change. For more Marxist-inspired critical theorists, economic change is a *sine qua non* of social reform. For those of a more Habermasian persuasion, the key to a society which serves universal rather than particular interests is the reclaiming of public life through rational debate, resulting in consensually derived forms of social co-ordination. These differences have important implications for some of the subsequent comparisons of PBM and critical research. Before turning to such a comparison, however, more needs to be said about critical theory as methodology.

The Methodology of Critical Research

While it is inappropriate to suggest that there is a set of techniques that distinguish critical research, several writers who represent either the more Marxist or the more Habermasian versions of critical theory provide similar accounts of its general procedure (Comstock, 1982; Fay, 1987, ch. 2; Geuss, 1981, ch. 3; Young, 1989, p. 41). Donald Comstock's account will be employed here because it provides the most detail, and is widely cited amongst critical researchers in education. He describes the approach as follows:

> Critical social research begins from the life problems of definite and particular social agents who may be individuals, groups, or classes that are oppressed by and alienated from social processes they maintain or create but do not control. Beginning from the practical problems of everyday existence it returns to that life with the aim of enlightening its subjects about unrecognized social constraints and possible courses of action by which they may liberate themselves. Its aim is enlightened self-knowledge and effective political action. Its method is dialogue, and its effect is to heighten its subjects' self-awareness of their collective potential as the active agents of history (p. 382).

The quote suggests that the starting point of critical inquiry is a practical problem revealed by the felt dissatisfaction of a group or individual. The explanation of the problem draws on both the actors' understandings of their situation and of the social structures which sustain and are sustained by those understandings. The concern with actors' subjective experience of their situation reveals the interpretivist dimension of critical inquiry. On this view, actors do not respond directly to external stimuli, but to their understandings and interpretations of such stimuli. Given the central role of ideology in critical theories, however, researchers cannot be satisfied with an interpretivist account, since insiders' understandings are likely to have been distorted by official, hegemonic accounts of social situations. The critical researcher demonstrates the falsity or unacceptability of these beliefs by showing either how they are incompatible with relevant evidence, how they lead actors to behave against their own interests, or how they were adopted for reasons that, if known, would be unacceptable to the actors themselves (Geuss, 1981, pp. 12–22). Ideology critique thus involves a dimension of traditional empiricist inquiry in its explanation of the origins of actors' understandings and in its evaluation of their accuracy.

This general framework for the analytic phases of a critical project can be seen in Paul Willis's (1977) landmark study entitled *Learning to labour: How working class kids get working class jobs*. In the first interpretive phase of the book, Willis presents an ethnography of the school and home experiences of two groups of youth in an inner city English comprehensive school. He shows how the "lads" celebrate the "manual" over the "mental" and endorse masculine stereotypes while shunning activities and attributes which could be construed as feminine. In contrast to the subculture of the lads, the ear'oles accept school values, including mental labour, and accordingly are more successful in school exams and job seeking. In the second more analytic and critical part of the book, Willis explains the beliefs of the lads in terms of the occupational conditions of their fathers; the subculture of the shop floor is both sexist and mistrustful of bookish learning. The lads' distrust of official rhetoric about the relationship between success at school and social mobility is based on a partially correct understanding of the ways in which schooling is irrelevant to their life chances. At the same time, however, the ethnography of the ear'oles shows how the lads misunderstand and exaggerate the irrelevance of schooling in ways which, in the long run, ensure that they reproduce rather than break free from the occupational and class location of their fathers. Willis's book is typically "critical" in its portrayal of the ways in which actors' understandings, and the social conditions which give rise to them, interact to produce and reproduce situations of dissatisfaction and disadvantage.

Critical research offers, in addition to critique, a constructive process of education and social action. The educative phase of a critical project involves the participants in a dialogue about the accuracy of the analysis, and about the desirability of alternative understandings and actions. In the process, the audience may accept the analysis, alter their understandings and plan to act to change the conditions which are, at least in part, causally involved in their unhappiness. According to Comstock (1982), the model of education which is appropriate to this process:

> is not the familiar one of formal schooling but, rather, a model of dialogue in which the critical researcher attempts to either problematize certain meanings, motives or values accepted by his or her subjects or to respond to issues which are already perceived by them as problematic (p. 385).

Comstock goes on to explain that if a critical theory is to serve as a catalyst to change, it must be comprehensible and acceptable to those whom it claims to be about. This implies either that researchers speak the participant's language while critiquing it, or that they employ a mutually achieved language that bridges the experience of researcher and practitioner. One way in which researchers betray the educative potential of critical research is by so transforming their subjects' experience that their accounts are unrecognisable to those whom they purport to be about. Another way is by imposing their analysis on the audience rather than involving them in a process of validation.

The educative phase of a critical project is followed by a social action phase, in which researchers and participants collaborate to redress the problems that were the subject of the analysis. Critical research is not idealist, for it recognises that changed understandings do not in themselves change material circumstances; action and ideas are interdependent in the process of understanding and changing the world (Fay, 1987, pp. 24–25).

Despite their allegiance to the critical tradition, many critical research projects in education stop short of the social action which is supposed to be one of the hallmarks of this tradition. For example, Willis's critical project did not offer his subjects either education or social action. After his extensive critical analysis, Willis adds some recommendations about "What to do on Monday morning" and an appendix which includes extracts of his conversation with the lads about his analysis. The dialogue was unlikely to have been educative for either party, since the abstract and highly theoretical nature of Willis's writing would have made his account of their lives barely recognisable to the lads, and Willis himself gives no indication that he construed the conversation as a serious test of his analysis. In addition, Willis's analysis does not show how the lads themselves could act to change their situation, and so lacks the emancipatory thrust of a critical theory. Instead, his recommendations for Monday morning are directed to others outside the research process, and as such reflect the traditional separation of research and political action.

Willis's failure to move beyond critique to social action is seen in many other purportedly critical studies. A search of the critically informed empirical literature in education showed that with the exception of the work of Ira Shor and Paulo Freire, most critical researchers do not carry their critique through to a stage of education and social action (Freire, 1985; Shor & Freire, 1987). Understanding why this might be the case, and its implications for PBM, is a major theme of the subsequent comparison of PBM and critical theory.

Comparing PBM and Critical Theory

The methodological framework outlined above suggests that both critical theory and PBM start with a problem situation, critically analyse the way actors understand and act in the situation and seek to resolve it consistently with an explicit normative theory. Both approaches, therefore, require theories of how actors may be understood as responsible agents and theories of education and social action. Beyond these procedural and interpretive similarities, however, there are pervasive theoretical differences between PBM and critical theory which are obvious at every stage of the process of problem selection, analysis and resolution.

In brief, the difference lies in the degree to which each methodology is associated with a particular social theory. Critical theorists employ substantive social theories which seek to explain how situations of oppression and injustice are created and sustained. Those who are influenced more by Marx, link social and educational problems to forces of capitalist production and class conflict; Habermasian critical theorists focus on how media, money and the administrative apparatus of the "system-world" take over the "life-world" by pre-empting opportunities for the relevant public to reach consensual decisions about the governance of institutions which are central to their lives. Each particular critical theory guides the analysis process, so that apparently diverse problems come to be understood in similar ways. The result is not only a claim to understand a specific educational problem, but a claim to understand, in general, why educational problems are so pervasive, serious and intractable.

PBM brings no such social theory to the problem-analysis process, so the question must be raised as to whether this jeopardises its goals of understanding and resolving educational problems. The position taken in PBM is that the complexity and diversity of social and educational problems, and the fallibility of our empirical and value claims, makes such an epistemological stance highly subject to error. Rather, PBM incorporates a normative theory of inquiry, which serves as a meta-level evaluative standard for appraising currently operative theories of the problem and for designing alternative theories. This meta-level theory asks: How are currently operative theories implicated in the problem? How would they need to be altered in order to resolve it? Who are the agents of learning in this situation? In PBM there are no *a priori* answers to these questions, only empirically derived answers for each problem situation.

This is not to imply that the approach of PBM is atheoretical. The theories that agents employ in the problem situation are evaluated against a normative theory, but, unlike critical theories, it only specifies microprocesses of inquiry, and is silent about the wider factors which may, in Habermasian terms, result in systematically distorted communication, or, in Argyris's terms, in unilateral rather than bilateral control of decisions about meanings, actions and normative standards. The origins of such distortions may be traceable, in particular cases, to economic factors, to the intrusion of the instrumentalist "system world" into the "life world", to features of individual development or psychology, or to any of a myriad of intrapersonal, organisational or socio-political factors. In PBM, no *a priori* judgements are made about either the necessity to trace the origins of maladaptive processes of inquiry beyond what is required for problem resolution (delimited by criteria of theoretical adequacy), or about the particular theory of origins which is likely to be more or less productive for understanding and resolving any particular educational problem. For example, it may be that hierarchical structures, and particular forms of accountability, are, contrary to what is frequently assumed, facilitative rather than inhibitive of interpersonal and organisational learning. The precise relationship of such structures to learning and critical dialogue is an empirical question which must be investigated in every case.

In the following pages, I trace the implications of these differences between PBM and critical theory for the achievement of their shared practical purposes. The discussion offers suggestions about why, in the opinions of several authors, critical theory has so far failed to deliver on its promises (Fay, 1987, pp. 211–215; Young, 1989, p. 68), and about how PBM does or can avoid the same difficulties. The discussion is organised around the stages of problem selection and problem analysis and resolution.

Comparing Problem Selection Procedures

In PBM, problems are selected where minimum conditions for their investigation and possible resolution are satisfied. Initially, there should be a match between the problem and the expertise and interests of the researcher, access to relevant data, and a commitment by some of the relevant stakeholders to its investigation. Such a commitment does not imply that researchers and participants have a shared theory of the problem, nor that a change process will be agreed upon. Such outcomes should be viewed as the possible results of, and not the preconditions for, inquiry. The audience, therefore, to which a problem analysis is addressed in PBM, and with whom decisions about its validity are made, are all those who have a stake in the problem situation, including both those who may be said to suffer under and benefit from the current arrangements.

In Marxist versions of critical research, by contrast, problems are selected by the social location of those who experience them. According to Comstock (1982), "Critical research is not *about* a social process but rather is *for* particular social groups—that represent progressive tendencies currently obscured and dominated" (pp. 379–380). While Comstock acknowledges some disagreement between critical theorists about precisely how such groups are identified, he suggests that those with progressive interests are those whose needs cannot be satisfied in current contexts of material and ideological domination. His examples of suitable candidates—working-class movements, trade unions, women's groups, poor people and minorities—suggest that educational administrators, policy-makers and perhaps even teachers would not be among them.

An obvious implication of this difference in problem selection criteria is that, unlike PBM, critical research is not open to investigation of problems experienced by more powerful and advantaged groups in society. Such research, it is argued, would become an exercise in technical problem-solving, rather than a normatively based examination of the way the powerful understand and attempt to resolve problems. The two case studies in this book belie this fear. The allegiance of the researcher employing PBM methodology is not to the party who experiences the problem, but to the enhancement of the problem-solving capacity of the system comprising all those experiencing and responsible for the problem. If the problem-solving capacity of that system is to be increased, then all these people must be treated as actual or potential research participants. The fact that initial negotiations to conduct the two case studies were conducted with the organisations' leaders did not prevent the researchers from gathering data from less powerful organisational members, and from using such data to develop an independent analysis of the problem which challenged many of the leaders' taken-for-granted assumptions about its nature and resolution. On other occasions, problems I have investigated have been initially identified by union delegates and not by the organisational leader. Regardless of the starting point, however, the consent and commitment of all key players is required to complete a process of critical inquiry into the problem and its possible solution.

It is not clear to what extent the same points about problem selection apply to Habermasian versions of critical theory. Young (1989) interprets Habermas as follows:

> The addressee of Habermas' theory is clearly universal, the oppressor *and* the oppressed. But the expectation that it is oppressed groups who will display the most immediate interest in enlightenment remains. In more recent work, Habermas appears to identify members of new social movements such as feminism and environmentalism as potentially fruitful addressees of critical theory, but his analysis by no means suggests that the working class is not an important addressee, or indeed, that elements of the elite cannot also be involved in enlightenment (p. 40).

Whether or not Habermasian versions of critical theory are more inclusive than more Marxist versions in their search for progressive groups, the difference with PBM remains because in the latter the normatively based problem-solving process requires the eventual involvement of all those who are part of the problem and part of the solution, whether or not they might be described as "progressive".

Comparing Procedures for Problem Analysis and Resolution

PBM's emphasis on inclusive processes of problem understanding and problem resolution contrasts with the methodological guidance given by Comstock to critical researchers. Having carefully selected a progressive group, they are advised to develop an explanation of the problem which draws on the understandings of *all* those involved (Comstock, 1982, p. 380). In an investigation of the homeless, for example, the understandings of city officials, slum landlords and property developers are to be investigated along with those of the homeless themselves and those who work on their behalf.

The inclusive nature of Comstock's advice about the analysis of the problem contrasts sharply with his advice about its resolution. His educational and political processes are directed exclusively at members of the progressive group (pp. 379–388). This seems strange given that the social theories that drive critical analyses nearly always attribute causal responsibility, through their control of relevant social structures, to non-progressive as well as to progressive groups. Paolo Freire, in similar fashion, sees the suffering of the oppressed as the result, at least in part, of social structures and practices which are controlled by the powerful, yet he excludes them from his emancipatory pedagogy (Freire, 1970, pp. 28 and 39). Paradoxically, while critical researchers locate the powerful in their analyses of problems, they exclude them from their solutions. The exclusion or bypassing of the powerful is counterproductive, given critical theorists' own claims that they are frequently partially responsible for the problem, through their direct or indirect control of the economic, political or communicative practices which sustain it. Unless revolutionary change is advocated or contemplated, social change requires the involvement of the powerful in a process of education and action designed to serve the critically examined interests of all. As Forester (1989) writes in the context of planning: "If planners ignore those in power, they assure their own powerlessness" (p. 27). It is puzzling that critical researchers develop sophisticated theories about the role of the powerful in social and educational problems and then bypass them in the process of change.

Critical researchers might defend this omission by arguing that the inclusiveness of PBM's approach to problem analysis and resolution naively presupposes that all those involved in a problem can work together to resolve it in ways that serve their common rather than their particular interests. Since the interests of the powerful are frequently in conflict with the interests of progressive groups, the political goals of critical inquiry might be compromised by such an inclusive approach. In the language of critical theory, the researcher cannot work for the oppressed and the oppressor simultaneously.

My response to this objection is to challenge the theory of interests that prejudges the possibility of including the powerful in the problem-solving process. A conflict of interests between two or more groups of individuals does not *necessarily* preclude collaborative effort towards a better world. Firstly, to say that *some* interests are in conflict is not to say that they are *all* in conflict. An assumption that they are, however, pre-empts a search for the common interests that could sustain a collaborative approach to the problem. After all, managers and union leaders do occasionally use their common interests in the survival of a firm, as a basis from which to resolve disputes about those interests which are more contested.

Secondly, the powerful, just like the powerless, could be mistaken about their interests and hence mistaken in their support of the status quo. As Geuss (1981) puts it, to say that current social arrangements work . . .

> "to the benefit and advantage" of the members of the dominant group means only that *in the given social system as then constituted* it is better to be a member of the dominant than of the oppressed group, i.e. that in *this* social order it is well to have as much normative power as possible. This in no way implies that the members of the dominant group are not themselves also massively frustrated, and also implies nothing about what social system they *would* prefer, if they had free choice (p. 87).

The assumption that a conflict of interests necessarily precludes collaboration is self-defeating, when the involvement of the powerful is a necessary condition for the non-violent resolution of the problem experienced by the powerless. In such conditions, it would be more productive for critical researchers to try to *create* conditions under which all those involved could test their perceptions of what is in their interests, than to prejudge their degree of conflict or commonality. This is the approach taken in PBM.

I turn now to a second difference between the approach of critical theory and PBM to problem analysis and resolution, which can also help explain the practical failure of many critical projects. The typical analyses produced in each approach differ in their relative emphases on structure and agency. In PBM, the focus is on the espoused theories and theories-in-use of participants and the way those theories may be implicated in the problem situation. This focus on the micro-processes of a problem situation contrasts with the more macro-level or structural analyses of many critical projects. While I do not want to set up a false structure–agency dichotomy, it is worth tracing the implications of these differing emphases for the achievement of the shared practical purposes of each approach.

Problem analyses which emphasise the role of structure are incomplete unless they show how those structures are mediated through the actions of particular individuals to cause the particular problem situation. For example, how precisely does the patterning of national examination practices prevent these teachers, in these schools, from introducing the curriculum they believe is necessary to reduce the alienation of their working-class students? Unless these connections are made, one cannot demonstrate the causal force of the particular social structure. A related issue concerns the extent to which agents can gain leverage over these structures. More Marxist versions of critical theory have been frequently criticised for their overly deterministic view of structure in which agents are left with little control over the problem because its more proximal causes are themselves tightly determined by wider economic and political forces. As Paul Willis (1977) has suggested, this type of critical research suffers from an immobilising circularity: "nothing can be done until the basic structures of society are changed, but the structures prevent us from making any changes" (p. 186).

One way in which this paralysis can be overcome is to produce analyses which identify precisely how particular structures are implicated in the problem situation and to involve those who can gain leverage over such structures in the problem-solving process. In order to follow this advice, critical researchers must turn from, or at least supplement, their more structural and holistic analyses, with the sort of detailed empirical work that can identify the possibilities for change that are open to particular agents in particular situations. Robert Young (1989) suggests why some critical researchers will reject this advice:

> The fact that many of the scholars who have turned to critical theory, then and now, have come to it from a background in the humanities ... has led to a very negative view of the value of empirical research, particularly quantitative research. This, in turn, has diminished the concreteness of their analyses somewhat. It is allied to a general characteristic of those who are attracted by the systematic scope of critical theory—a tendency to keep analysis at a highly abstract level (p. 69).

In summary, analyses which do not connect relevant structures to the actions and understandings of particular agents are causally incomplete, and fail to reveal the concrete possibilities for reconstruction that are open to those whom the critical theory addresses.

The opposite imbalance of an overemphasis on agency may also produce problem analyses which are incomplete and ineffective. Habermasian versions of critical theory have been criticised for exaggerating the capacity of individuals to transform themselves and their situation (Fay, 1987, ch. 8; Young, 1989, p. 62). This type of critical theory and PBM may both be dismissed as utopian if the possibilities for educational and social change are not seen in the context of the enormous internal and external barriers to such projects. An adequate problem analysis must recognise both the possibilities for change and the obstacles in its way, and treat both as hypotheses in need of empirical test. The strong emphases in PBM on empirical inquiry and on intervention provides safeguards against utopian claims about the possibilities for change. There is nothing like trying to do it, as demonstrated in Part II of this book, to realise the enormous complexities and difficulties involved in normatively based and democratically conducted interpersonal and organisational change.

Finally, on this theme of imbalance in problem analyses, I would like to argue that given that an analysis is imbalanced, it is preferable that it exaggerate the causal responsibilities and action possibilities of agents, rather than exaggerate the way the situation is constrained by wider structures. The former imbalance is correctable through the feedback that comes with attempts to act from the analysis. The latter imbalance is far less likely to be detected, because its structural determinism is less likely to suggest how its own claims can be tested in the particular context under investigation. There is a practical payoff, in other words, in formulating analyses in ways that encourage agents to confront constraints and thus test their and our theories about what is and is not possible.

My final set of explanations for the limited practical contribution of critical research concerns a much more practical but no less important set of explanations. Since most critical projects in education have stopped short of the educative and action phases of critical inquiry, there is a very limited database from which to develop a theory of these processes. Some possible explanations for this state of affairs have already been discussed, in terms of critical theorists' failure to engage the powerful and to produce analyses which act as catalysts for local change efforts. Other reasons can probably be found in the disciplinary background and training of critical researchers, and in academic disincentives to engage in long-term collaborative change projects which, by definition, they cannot unilaterally control. In addition, critical researchers have been unwilling to mine the theoretical and practical knowledge of change which is to be found in traditions like psychotherapy and organisational development, possibly because they have been stereotyped as too psychological or as having the wrong politics. If critical researchers are to take their own ideals seriously, they must engage in the messy, stressful and practical business of change, alongside those they purport to help, and thereby develop both their own intervention skills and the theory of education and social action that informs critical theory.

In summary, I have argued that the theoretical differences between critical research and PBM have implications for the way research problems are selected, formulated and resolved. The politics of critical research leads investigators to ally themselves with what Comstock calls "progressive groups" and to treat their interests as in conflict with and thwarted by those of more powerful groups. In PBM, by contrast, the allegiance of the researcher is to a democratic problem-solving process, not to any of the interest groups involved, whether relatively powerful or powerless. Such allegiance requires the researcher to try and create conditions in which all parties perceive sufficient commonality of interest to enter into and sustain the problem-solving process.

Habermas's Communicative Action and Critical Dialogue Compared

The methodologies of critical theory and PBM are simultaneously critical and collaborative. In PBM, the achievement of a truly critical and collaborative relationship is guided by the theory and practice of critical dialogue, based on the model of interpersonal effectiveness which Argyris calls Model Two (Argyris, 1983). In contrast, many critical theorists in education have turned to the work of Jurgen Habermas in their search for a model of democratic emancipatory dialogue (Burbules, 1991; Marshall & Peters, 1985). To my knowledge, Argyris and Habermas have not seriously studied each other's work. The question I wish to pursue, therefore, is what resources each approach can offer the other in achieving their goal of warranted agreement about the analysis and resolution of social and educational problems.

Habermas's interest in interaction stems from his belief that formal features of what he calls communicative action presuppose a concept of rationality that overcomes the limitations of the much narrower instrumental rationality discussed in my earlier introduction to Habermasian versions of critical theory. If his argument succeeds, then he shows how we are all committed through our status as competent speakers to the rational redemption of claims about the world, about ourselves and about what is right (Habermas, 1984, pp. 306–307). Rationality, as a communicative ideal, is not something that we accept or reject, attribute to a particular culture, or subject to similar relativist criticism. It is given through our language; in Habermas's phrase, it is a pragmatic precondition of argument.

Habermas's theory of rationality can offer PBM a strong foundation for the normative basis of critical dialogue. Precisely what practical difference this would make, however, is unclear, because Habermas recognises that this rational ideal is counterfactual, i.e. that it remains for the most part unrealised in our communicative practices. Habermas's foremost concern is the elaboration and defense of this ideal of rationality, not the investigation of the empirical detail of how it may or may not be instantiated in particular contexts. In contrast, it is precisely these concrete practices that must be worked out if PBM and critical dialogue are to contribute to the improvement of educational practice. In short, my position is that while Habermas's theory can contribute to the normative justification of PBM, it offers far less to the working out of how these ideals can be attained in the face of powerful external and internal constraints. In the following pages, I expand these arguments through a closer comparison of PBM and Habermas's writing on communicative action.

Habermas's expanded notion of rationality offers PBM a defence of its claim that disputes about norms, as well as about empirical matters, are resolvable through mutual disclosure and critique of reasons. He argues that the possibility of rationally resolving such disagreements is given by virtue of the formal pragmatics of speech. By this he means that every speech act undertaken in a context of communicative action embodies three different types of validity claim. The very act of making such claims presupposes the ability, if required, to redeem them through the provisions of reasons, and it is this that guarantees a consensual co-ordination of action and understanding between speaker and hearer. Habermas (1990) summarises these ideas in the following explanation of communicative action:

> I call interactions communicative when the participants coordinate their plans of action consensually, with the agreement reached at any point being evaluated in terms of the intersubjective recognition of validity claims. In cases where agreement is reached through explicit linguistic processes, the actors make three different claims to validity in their speech acts as they come to agreement with one another about something. Those claims are claims to truth, claims to rightness, and to truthfulness, according to whether the speaker refers to something in the objective world (as the totality of existing states of affairs), to something in the shared social world (as the totality of the legitimately regulated interpersonal relationships of a social group), or to something in his own subjective world (as the totality of experiences to which one has privileged access) (p. 58).

In short, Habermas is arguing that the rationality of claims about the world, about ourselves and about what is right are all redeemed through appeal to relevant argument and evidence.

A speaker's ability to demonstrate the rationality of these claims is usually taken for granted, because everyday communication takes place against a background consensus. For example, when someone states in a meeting that time is running short, we usually accept the claim because we know what the time is, when the meeting usually finishes and what else is still to be discussed. If, however, we want participants to reconsider the conventions that have developed about the use of time, we might challenge such a claim. If such challenges cannot be met immediately by appeal to relevant evidence or norms, then the speakers have recourse to a meta-level discourse in which the disputed claims, including the language in which they are expressed, can be critiqued. Disputes are adjudicated via discourse when evidence is indeterminate and initial arguments fail to compel (Dews, 1986, p. 162).

Unlike Argyris, who is merely claiming that dispute resolution ought to proceed rationally, since such procedures serve learning and problem-solving better than other non-rational ones, Habermas (1984) is making the much stronger normative claim that it must proceed rationally, since "no one would enter into moral argumentation if he did not start from the strong presupposition that a grounded consensus could in principle be achieved among those involved" (p. 19). This difference can be better understood if it is put into the context of Habermas's quest for a universal grounding of his broadened concept of rationality, so that it escapes the charge that it is the product of, or only applicable to, a particular reading of Western culture.

At the same time as defending his normative ideal, Habermas (1990) is clear that it is counterfactual, that is, that discourse is seldom conducted in the manner suggested by the ideal. In discussing practical discourse (the form of discourse in which disputes about normative claims are resolved), he writes:

> Like all argumentation, practical discourses resemble islands threatened with inundation in a sea of practice where the pattern of consensual conflict resolution is by no means the dominant one. The means of reaching agreement are repeatedly thrust aside by the instruments of force. Hence, action that is oriented towards ethical principles has to accommodate itself to imperatives that flow not from principles but from strategic necessities (p. 106).

One of the forces that precludes discourse is that of strategic action. Habermas describes the difference between strategic and communicative action as follows:

> Whereas in strategic action one actor seeks to *influence* the behavior of another by means of the threat of sanctions or the prospect of gratification in order to *cause* the interaction to continue as the first actor desires, in communicative action one actor seeks *rationally* to *motivate* another by relying on the illocutionary binding/bonding effect (*Bindungseffekt*) of the offer contained in his speech act (p. 58).

It is in the relationship between strategic and communicative action or, in Argyris's language, between the ideal of Model Two and the actuality of Model One, that the differing purposes of PBM and Habermas become most obvious and important. Habermas's concern with strategic action is to show how it does not undermine his appeal to language as the basis of his communicative norms (Habermas, 1984, pp. 288–294). The success or failure of his argument is only indirectly relevant to PBM, however, for the latter's concern is with how strategic action interferes with the practice of those norms, with the consequences of strategic action for problem resolution and with the replacement of strategic with discursive practices. Though language and argumentation may provide the possibility of rationally redeeming all disputed validity claims, PBM research must focus quite specifically on why this possibility is so seldom realised. It is as if speakers are saying, "Yes, we do adhere to the ideal of a free rational consensus, but believe that in this context, the ideal is not applicable, because others are acting strategically, the organisation won't allow it, or I can't be beaten on this one." At other times, speakers do not set aside these ideals; they are simply unaware of the ways in which they act inconsistently with them. Habermas (1990) explains his interest, or rather disinterest, in the "choice" between strategic and communicative action as follows:

> The possibility of *choosing* between communicative and strategic action exists only abstractly; it exists only for someone who takes the contingent perspective of an individual actor. From the perspective of the lifeworld to which the actor belongs, these modes of action are not matters of free choice. The symbolic structures of every lifeworld are reproduced through . . . processes [which] operate only in the medium of action oriented toward reaching an understanding. . . . That is why they, as individuals, have a choice between communicative and strategic action only in an abstract sense, i.e., in individual cases. They do not have the option of a long-term absence from contexts of action oriented toward reaching an understanding. That would mean regressing to the monadic isolation of strategic action or schizophrenia and suicide. In the long run, such absence is self-destructive (p. 102).

Once again there is no incompatibility between Habermas and PBM, only a very different focus. It is precisely what Habermas calls the "individual cases" that concern PBM, for it is these that may need to be critiqued and reconstructed in order to resolve particular problems. Appeals to a norm which have already been rejected hold little sway in such cases. What is needed is an argument that shows how strategic action is counterproductive in the speaker's own terms and a demonstration of how discourse is both possible, and more satisfying, than strategic interaction. The problem focus of PBM provides a context for debating the consequences of strategic action and for evaluating any claims for the desirability of the more rational and consensual alternative. Beyond this, critical dialogue specifies the specific communicative moves that may need to be taught before speakers can test for themselves the consequences of abandoning their strategic choice.

Analysing Actual Speech: A Research Example

Throughout this volume, and in the empirical research on which it builds, I have been guided by the communicative theories and practice of Argyris and not those of Habermas. In part, this reflects my own learning history, since I studied the latter's work in only the last five years. More importantly, however, it reflects the differences in their purposes which have been stressed in this discussion. Argyris's writing offered a normative theory of interpersonal and organisational learning which was grounded in practice and a methodology by which it could be applied to empirical data. Habermas's theory of communicative action is concerned to defend an ideal of rationality, founded in the universal presuppositions of speech, and this ideal cannot be directly applied to everyday situations. As Habermas himself says, formal pragmatics, which is concerned with the identification of these presuppositions, is "hopelessly removed from actual language use" (Habermas, 1984, p. 328). Robert Young (1989) summarises the difference between formal pragmatics and empirical pragmatics, which is concerned with the communicative practice of everyday life, as follows: ". . . the theory of universal pragmatics [. . .] is an abstract or general theory developed against the background of an undifferentiated and notional social context" (p. 105). It is intended only as a guide for empirical developments, and has been mistakenly employed, including by myself, as a standard against which actual speech can be directly assessed (Robinson, 1989). Empirical pragmatics, by contrast, is the study of actual utterances in specific, differentiated social contexts (Young, 1989, p. 195). Young goes on to say that this would involve analysing sequences of naturally occurring dialogue, investigating how the background of speakers influences their interpretations and recognising instances of distorted (strategic) communication.

Given the similarity between the theoretical frameworks of Habermas and Argyris, I wish, in the remainder of this chapter, to discuss an example of how such an analysis proceeds in the latter's framework as a way of stimulating more empirical analyses of the type that Young calls for, in the hope that such empirical work may eventually improve the practical payoff of critical research.

THE RESEARCH CONTEXT

The context of this analysis is a negotiation between a university researcher and an elementary school principal about the validity of the former's account of the latter's administrative practice. In their 1987 AERA conference paper "Negotiating Organizational Reality: A Case Study of Mutual Validation of Research Outcomes", the two parties include an extract of their negotiation to illustrate their collaborative methodology (Anderson & Kelley, 1987). The extract is reproduced below (Table 11.1), and then analysed to illustrate how Argyris's framework (Figure 3.1) may be used to judge the degree to which the speakers achieve a rational consensus about their interpretation of the research data. I will argue that, despite the authors' agreement about their conclusions, two of the three are not warranted because they have not been subject to the type of test which is associated with a search for valid information. (Before proceeding further, I want to acknowledge both that Anderson and Kelley's paper is one of the few I could locate in which discussion of researcher–practitioner dialogue is empirically illustrated, and that the analysis has benefited from correspondence with Anderson.)

Prior to the dialogue extract, Anderson and Kelley had disagreed over the former's characterisation of Kelley's environment as "unsafe"—a metaphor he had selected to capture the principal's frequent references to "tremendous stress, sleepless nights, potential lawsuits, and constant strategizing" (p. 10). In discussion, Kelley had claimed that her relations with her staff, parents and immediate boss were much more harmonious and cohesive than was implied by the "unsafe" metaphor. Reanalysis of his data convinced Anderson that there was evidence to support a more benign "family" metaphor, and he acknowledged that "he had come to the study with a bias toward a conflict paradigm of social reality in which conflicts of interest are resolved through the exercise of political power" (p. 12).

Anderson then took his revised analysis back to the principal, to check whether both the "unsafe" and the "family" metaphors could accurately account for how she approached administrative decision-making. The conversation presented in Table 11.1 was part of their attempt to reconcile the two apparently conflicting metaphors.

How well-grounded are these conclusions? Have they been subject to the force of argument and counter-argument? What is the quality of the consensus that has been reached between principal and researcher? In Argyris's terms, to what extent do the speakers' moves reflect a commitment to valid information rather than to winning, that is, to persuading the other of the correctness of their views?

TABLE 11.1. *A Negotiation of Research Conclusions* (Anderson & Kelley, 1987)

Principal: Sometimes you have a joyous thing—you've achieved some success. But next phone call could just make you go down again. I don't know, you never quite maintain a high in this job. And I think that's affected my personality somewhat. I was always an optimist and always a pretty high type of person, lots
5 of energy. And I find that now I get pretty worn out a lot, and things don't thrill me as much as they used to.

Researcher: I feel like the unsafe place metaphor is operating but I'm not sure how it fits in, and what you seem to be saying is yes, the stress is there and the
10 . . .

Principal: Well, look, a part of unsafe is that there are certain things I would never say to my staff that I might say to a couple of my colleagues because you need to let it all out.
15

Researcher: And that's stress provoking.

Principal: Well sure, I mean it's stressful that you can't say it to them. Sometimes I'll get so angry and I'll think, "just do your job and quit balking" or something like
20 that, whereas I can say that to a couple of my principal colleagues and it's a way I let it out so when I come back to these folks, I can deal with them in the way I know I need to deal with them.

Researcher: Is this partly a hierarchy or accountability issue where you can't just
25 say to your staff "go ahead and do what you want", because that also reflects on you and the school?

Principal: Oh yes, like the state evaluation thing. If my staff hadn't followed through on that, that would have been me, that would have been my fault if they
30 hadn't. I mean not that it's said, "that's your fault" but I know if they hadn't pulled through on that . . .

Researcher: So bureaucracy is a fact of life; hierarchy is a fact of life and there is a certain amount of pressure that comes from being part of a hierarchy.
35

Principal: Right.

Researcher: So in a sense that kind of stress is inherent because of the accountability that occurs at each level. Now it seems that the way this district
40 deals with that—and rather effectively—is, in order to cut down on the impersonality of the bureaucracy and accountability, to create a sense of cohesiveness at each level and to effectively but selectively funnel information up and down the hierarchy.

45 *Principal*: Well, I don't think the funnelling of the information would make it less.

Researcher: Yeah: you're right. That doesn't make sense. How about this? At each of these levels, that idea of the family or unit or team metaphor is used to build the cohesiveness, but yet at the same time they are power bases or interest
50 groups within the community?

The Dialogue Analysis

The warrant for each of the three conclusions is examined in terms of the moves that are made to ground the claims in data, to provide reasons, to raise alternatives and to check with the other party (Table 11.2). Instances of such moves, along with line references (l), are listed in the second column; significant omissions in the third.

The first conclusion involves the claim that the unsafe metaphor is applicable to parts of Kelley's job. What is at stake here is simply the accuracy of a description, and the speakers establish its validity by providing illustrations and agreeing, presumably based on a shared implicit theory of stress, that such circumstances can be thus described.

The second conclusion requires a more complex type of warrant because it involves an explanatory theory. Despite the principal's agreement to and illustration of the theory, I suggest in Table 11.2 that its validity has not been established because competing explanations have been overlooked. Instead of testing his hypothesis about bureaucracy and accountability by inquiring about the situations which trigger the principal's angry feelings, Anderson reaches for his political theory and links her stress to bureaucratic demands which she is forced to pass on to her staff. While the principal confirms and illustrates this argument, such a confirmation is an inadequate test of Anderson's assumption that her stress was caused by external demands. First, the stress may be related to the way the principal deals with these demands, rather than being an inevitable consequence of them. For example, a principal who is unable to share the reactions she has to staff's reluctance to carry out mandated activities would be far more stressed than one who is able to disclose these dilemmas to her staff and seek their help in resolving them. The principal's stress may be a result of an *interaction* between the external demands and her ways of coping with them, rather than of the external demands themselves. A second stronger test of the relationship between stress and external demands would involve a search for different antecedents of similar stress reactions. If suppressed angry feelings, which are experienced as stressful, are also triggered by internally generated demands (e.g. staff's failure to meet the principal's own standards), then this evidence would tend to disconfirm the strong link Anderson claims between hierarchical accountability and the experience of stress and of being unsafe. Anderson's failure to seek disconfirmation of this analysis and his ignoring of the principal's account of her coping strategies (lines 19–22) seriously reduces the credibility of his attempt to overcome his acknowledged bias towards political theories of organisations.

Although Anderson offers some reasons to support his third conclusion about the relevance of the family metaphor and about its

TABLE 11.2. *The Analysis of Validity of Negotiated Conclusions*

Conclusions reached	*Warrants provided*	*Warrants not provided*	*Validity of conclusion*
1. Parts of the principal's job are unsafe and stressful.	• has to suppress angry feelings towards staff (l. 12). • such suppression is stress-provoking (l. 18). • expressing thoughts to others relieves stress (l. 20). • P and R agree on appropriateness of metaphors (l. 18).		• established.
2. Hierarchy and bureaucracy are a source of the principal's stress.	• because they prevent P from allowing staff to go ahead as they wish (l. 24). • R illustrates how hierarchical accountability can be stressful for her (l. 28). • P and R agree on explanation (l. 36).	• assumption that P would wish staff to do what they want not checked. • neither P nor R test their explanation of stress against competing hypotheses.	• not established.
3. The family metaphor is used to reduce impersonality and create cohesiveness at each level.	• one-way cohesiveness is established through funnelling information (l. 42). • R and P agree to reject funnelling idea (l. 47). • levels function as both family units and power bases (l. 49).	• evidence for funnelling not provided. • reason for rejection not provided. • no reasons or evidence provided.	• not established.

Note: Numbers in parentheses refer to dialogue lines in Table 11.1.

operations, those reasons are themselves unsupported by evidence. He indicates later in the paper that a consensus was reached on the applicability of the metaphor, but the basis of the agreement cannot be derived from the dialogue.

A number of points are worth noting about this dialogue and its analysis. First, it occurs in a context which one would expect to be highly favourable to communicative action. Anderson has publicly committed himself to a process of "mutual validation of research outcomes", he is aware through prior reflection of his propensity to impose his political perspective and he asks for the principal's help in revisiting the data. Despite this, he and the principal fail to "mutually validate" two of their three conclusions. Argyris's theories about interpersonal and organisational learning suggest that such failures are not casual or random mistakes. They are the result of a highly skilled capacity to act strategically, i.e. to persuade others of our views without subjecting them to adequate test, despite the commitment and the capacity to act otherwise. Examples like this one suggest to me the importance of conducting empirical analyses of what is involved in initiating communicative action, and of educating ourselves and others in its employment.

A second methodological point is suggested by the fact that communicative and strategic action cannot be discriminated by the mere offering and acceptance or rejection of reasons. Our empirical analyses need to incorporate judgements about the adequacy of those reasons, because the reasons given to support the second and third conclusions are far from compelling. Argyris's concept of testing, and his analyses of what is involved, including the search for disconfirmation and competing examples, are a useful basis for such discrimination. If Habermas's concepts of consensually co-ordinated action and rationality are to do the practical work in education and educational research that many authors wish for, it is time to build on the theoretical base that he provides with a programme of empirical research which seeks to show how particular forms of dialogue are problematic, and how more normatively desirable forms are both possible and more effective in those same contexts.

Summary

Critical research poses a substantial challenge to PBM, because it shares its practical purposes and pursues them through an apparently similar methodology. Both begin with a problem situation, provide a causal analysis of the problem which incorporates the understandings of those who experience it, critique those understandings and offer a normatively based process of education and action to address the problem. Beyond these similarities, a more detailed comparison between PBM and critical research requires a distinction between the more Marxist and more Habermasian versions, because the latter's focus on rational consensus through dialogue makes it much more like PBM than the former's more structural and macro analyses of problem situations and of social change.

After an initial exposition of these two versions of critical theory and its methodology, the chapter addressed the question of why critical research in education had so far failed to deliver on its practical promise. It was answered through discussion of the problem selection, analysis and resolution processes associated with critical theory and through a comparison of these procedures with those in PBM. The search for progressive groups, the theory of interests that leads to the exclusion of the powerful, and analyses that emphasise abstract structural determinants of problems to the neglect of more manipulable factors were all seen as possible explanations. In addition, the failure of many critical researchers to carry their projects through to the educative and social action phases, together with the neglect of theory and research on change that originates in other traditions, were also seen as impoverishing the change potential of critical research.

The second half of the chapter explored the resources which Habermas could provide for the normative and dialogical aspects of PBM. While Habermas's theory of rationality and communicative action is very similar to that which underlies critical dialogue, Habermas's concern with normative foundations is very different from PBM's concern with the improvement of practice. While his universal pragmatics yields normative guidelines which are very similar to those of Argyris's Model Two, they cannot be directly applied to everyday speech, and so empirical applications are dependent on the development of an empirical pragmatics which works in complex empirical contexts. The chapter concludes with a small contribution to such a research programme by showing how the values and strategies of critical dialogue can be applied to actual speech to reach judgements about the presence of strategic communication and about its consequences for validity claims.

References

Anderson, G. (1989). Critical ethnography in education: Origins, current status and new directions. *Review of Educational Research*, **59** (3), 249–270.

Anderson, G. & Kelley, D. M. (1987). Negotiating organizational reality: A case study of mutual validation of research outcomes. Paper presented at the 1987 Annual Meeting of the American Educational Research Association, Washington D.C.

Argyris, C. (1982). *Reasoning, learning and action*. San Francisco: Jossey-Bass.

Bowles, S. & Gintis, H. (1970). *Schooling in capitalist America*. New York: Basic Books.

Braybrooke, D. (1987). *Philosophy of social science*. Englewood Cliffs, NJ: Prentice-Hall.

Bredo, E. & Feinberg, W. (Eds.) (1982). *Knowledge and values in social and educational research*. Philadelphia: Temple University Press.

Burbules, N. C. (1991). Forms of ideology critique: a pedagogical perspective. *International Journal of Qualitative Studies in Education*, **5** (1), 7–17.

Carr, W. & Kemmis, S. (1986). *Becoming critical: Education, knowledge and action research*. London: Falmer Press.

Comstock, D. (1982). A method for critical research. In E. Bredo and W. Feinberg (Eds.), *Knowledge and values in educational and social research*, pp. 370–390. Philadelphia: Temple University Press.

Dews, P. (Ed.) (1986). *Habermas: Autonomy and solidarity: Interviews*. London: Verso.

Ewert, G. D. (1991). Habermas and education: A comprehensive overview of the influence of Habermas in educational literature. *Review of Educational Research*, **61** (3), 345–378.

Fay, B. (1987). *Critical social science*. Ithaca, NY: Cornell University Press.

Forester, J. (1989). *Planning in the face of power*. Berkeley: University of California Press.

Freire, P. (1970). *Pedagogy of the oppressed*. New York: Herder and Herder.

Freire, P. (1985). *The politics of education: Culture, power and liberation*. South Hadley, MA: Bergin and Harvey.

Geuss, R. (1981). *The idea of a critical theory: Habermas and the Frankfurt school*. Cambridge: Cambridge University Press.

Habermas, J. (1984). *The theory of communicative action* (*Vol. 1. Reason and the rationalization of society*) (T. McCarthy, trans.). Boston: Beacon Press. (Original work published 1981.)

Habermas, J. (1990). *Moral consciousness and communicative action*. (C. Lenhardt & S. Weber-Nicholson, trans.). Cambridge: MIT Press. (Original work published 1983.)

Lakomski, G. (1988). Critical theory. In J. P. Keeves (Ed.), *Educational research methodology and measurement: An international handbook*, pp. 54–59. Oxford: Pergamon Press.

Marshall, J. & Peters, M. (1985). Evaluation and education: The ideal learning community. *Policy Sciences*, **18**, 263–288.

Robinson, V. M. J. (1989). The nature and conduct of a critical dialogue. *New Zealand Journal of Educational Studies*, **24** (2), 175–187.

Shor, I. & Freire, P. (1987). *A pedagogy for liberation: Dialogues on transforming education*. Basingstoke: Macmillan.

Thompson, J. B. & Held, D. (Eds.) (1982). *Habermas' critical debates*. London: Macmillan.

Walker, J. C. (1985). Materialist pragmatism and sociology of education. *British Journal of Sociology of Education* **6** (1), 55–74.

Willis, P. (1977). *Learning to labour*. Farnborough: Saxon House.

Young, R. (1989). *A critical theory of education: Habermas and our children's future*. New York: Harvester Wheatsheaf.

PART IV

Conclusions

12

Obstacles and Opportunities for Problem-based Methodology

From time to time, a highly negative assessment is made of the practical contribution of educational research. Among the latest is that of Robert Young (1989) who writes:

> Never in the history of science has a research community produced so many "results" or "findings" with so little effect as have educational researchers since the advent of the electronic computer—one would have thought that there would, by now, be a thirst for critique, rather than a burgeoning of defensive claims about research's effectiveness (p. 137).

Whether or not such an assessment is a cause for concern on the part of educational researchers depends in part on the causal theory that they invoke. If researchers attribute the poor contribution to the fact that they are minor players on the education stage, that many of the factors relevant to problem resolution are outside the control of either educators or researchers, or to the fact that research findings are only one of a host of factors that determine educational policy and practice, Young's evaluation will trigger little more than resigned acceptance. If, on the other hand, such evaluations are seen as the result, at least in part, of the methodological choices that researchers themselves make, then it is more likely to trigger a critical rethinking of the way they do research, or at least of the way they do research that is intended to influence practice.

In this volume I have urged the second type of response, by arguing that a major reason that educational research has had so little influence on practice is that much of it is seriously mismatched to the subject matter it seeks to illuminate and influence. Many standards of methodological adequacy bear little relationship to, or are even contrary to, the requirements of practice. As a consequence, researchers have relied on influence processes external to their methodology to bridge the gulf between their research designs and the practice context. The literature on dissemination suggests that this strategy is only likely to be successful when research findings are broadly consistent with current ways of formulating the problem; when the theories of researchers and practitioners are radically different, research findings are likely to be rejected.

This volume has described a methodology for doing research for the improvement of educational practice, through collaborative and critical processes of problem formulation and resolution. First, problem-based methodology requires that educational problems be understood as the intended or unintended consequences of prior problem-solving efforts. Researchers should understand them by understanding the constraints that led relevant actors to enact the practices which we now judge to be problematic. Much research has failed to influence educational problems because it has separated problematic practices from the pre-theorised problem-solving processes that gave rise to them and which render them sensible to those who engage in them. Once practice is understood in this way, the theorising and reasoning of practitioners becomes a key to understanding what sustains problematic practice. Problem-based methodology provides a way of uncovering, evaluating and, if necessary, reconstructing these theories of action.

Second, taking practice seriously means accepting its requirement for normative theorising. Knowledge about what is desirable is as important to practitioners as knowledge about what is the case, and methodologies that eschew the former are seriously mismatched to practice requirements. PBM offers a set of normative standards which suit its problem understanding and problem resolution purposes, and at the same time recognises, through its improvability criterion, the fallibility of such judgements.

Third, PBM involves relevant practitioners in a dialogical process that is simultaneously critical and collaborative. Some educational problems persist because various stakeholders have been unable to agree on a shared theory of the problem and of its resolution. PBM conceptualises such disputes, including those between researchers and practitioners, as theory competition, and seeks to resolve them through critical dialogue with those involved. When the theorising of practitioners is taken seriously, their objections are not treated as obstacles to be overcome in a subsequent dissemination phase, but as a theoretically informed challenge to others' theories of the problem and of how to resolve it. These social relations of inquiry enhance theoretical adequacy through the encouragement of mutual critique and foster the intellectual commitment required to drive the problem resolution process.

The Significance of a Problem Focus

Some may doubt that there is much that is new in the problem focus of PBM, since any research can be loosely construed as problem-focused, by, for example, describing the research hypothesis or question as the problem to be investigated. It is worth, therefore, restating that in PBM a problem is a set of constraints on what counts as a solution plus the demand that a solution be found. When the goal or desired state is separated from the constraint structure that specifies what counts as meeting it, a simplistic problem formulation results which is likely to be rejected by practitioners. For example, a hypothetical school leader's problem is not how to be democratic, but how to be democratic given constraints of time, school size, politics and her own skills. A hypothetical school district's problem is not how to hold teachers accountable, but how to do so consistently with other values of professionalism and teachers' rights. These problems are such, precisely because they have multiple criteria of solution adequacy which are frequently in conflict, require tradeoffs and place those who would solve them in various types of dilemma. One can understand why practitioners would reject research that simplifies or bypasses this constraint structure, for its applicability to their more complex context is unknown. This is not to say that such research should not be conducted. Research problems arise in the context of academic as well as practical theorising, and research on the former, which frequently requires a simplification of the practice context, is an essential resource for the critique and reconstruction of theories of practice. What PBM makes clear, however, is why research which addresses one type of problem is not easily connected to the other, and the methodological implications of choosing the practical path.

The significance of the problem focus of PBM cannot be overstated. It helps resolve the motivational, normative and ethical problems that beset research which is intendedly critical and collaborative. Regarding the first motivational issue, an obvious question for critical researchers is why should or would practitioners want to co-operate ? After all, there is nothing particularly pleasant about being the subject of critique, no matter how well disguised in structuralist abstraction, and without a problem focus it is not at all clear how the interests of the practitioner, as opposed to those of the researcher, are to be served. PBM addresses the motivational problem through its commitment to address, and if possible resolve, a problem which practitioners experience. The pain of critique and the effort of rethinking and redesigning established practices is worth undertaking, not for its own sake, but because it offers the promise of resolving a problem which there is a sincere, public and shared desire to address.

A problem focus also helps resolve the question of what normative and evaluative standards are to regulate the research process. To agree to enter a problem-solving process is to agree to those standards of theoretical adequacy that promote those purposes. I argued in Chapter

2 that those standards should be explanatory accuracy, effectiveness, coherence and improvability, and it would be hard to deny their importance, in principle, to solving sets of ill-structured problems. This is not to say that these criteria might not be rejected in particular cases, as we discovered in Northern Grammar, when they conflicted with other normative criteria, such as emotional comfort, that were more deeply held than those related to inquiry. Agreement on standards of theoretical adequacy does not preclude disagreement on the application of those standards. Once again, however, a shared problem focus provides a context in which many such disagreements can be resolved, because shared knowledge of the problem context provides significant constraints on the number of plausible interpretations. Disputes about the likely effectiveness of different action steps, for example, may be resolved by appealing to shared past experience of the situation or by pointing to inconsistencies between one of the claims and previously agreed causal analyses. As the scope of a shared problem analysis widens, so it progressively reduces the degrees of freedom available for subsequent interpretations of the problem.

A third advantage of a problem focus is that it provides some ethical safeguards for the exercise of critique. An ever-present critical stance is neither possible nor desirable in research which seeks to engage the subjects of that critique. Critique that is any more than an abstract intellectual exercise, or entirely other-directed, is both emotionally and cognitively taxing. Practitioners are locked into webs of practice that are extremely difficult to rethink and redesign, and to invite them to do so entails some considerable obligation on the part of the researcher to see it through and to try to ensure that the process leaves those involved no worse off, at least, than when they started. A problem focus limits the scope of critique and focuses effort and energy in a way that is more likely to yield positive outcomes. The critique is justified by its contribution to the understanding and resolution of a problem situation with which the practitioner is thoroughly familiar; it is this familiarity that enables the practitioner to hold the researcher accountable for his or her contribution to understanding and resolving the problem situation.

Some researchers may object that these ethical concerns arise because I have portrayed critique in too personalised a manner; that it is patterned relationships and structures which are the subject of critique, and that these exist independently of the individuals who happen to be enacting them at the time. This type of objection reminds me of a critical ethnography that I read recently in which the author, having described in some detail the way teachers and students in a particular high school interacted to produce a curriculum and a style of pedagogy that reproduced rather than broke down class and ethnic difference, then professed not to be criticising the individuals she had studied in such detail, but to be criticising the structures which patterned their interactions and determined their outcomes. The author is clearly in something of an ethical dilemma, because to deny the teachers any responsibility for the patterns she observed is to depersonalise them in ways that they would probably find neither acceptable nor credible; yet to attribute responsibility to them is to criticise them in ways that she clearly felt was a breach of faith. The example points up the dilemma that is faced by those who employ structuralist social theories and then seek to make constructive contributions to the situations that are the subject of their critique (Strike, 1989). Critique must be personalised if it is to identify how particular agents have acted to sustain a problem situation and how they could act to resolve it. Such personalisation does not imply that those individuals' actions are not themselves responses to particular structural conditions; it does imply the possibility that individuals could have acted otherwise, and that such a possibility is worth testing. Once critique is personalised, then the ethical and political qualities of the researcher–practitioner relationship became highly significant in the success of a project. The advantage of a problem focus is that the critique is anticipated, agreed to and mutually regulated in terms of its contribution to the understanding and resolution of the focus problem.

Versions of critical dialogue, one of the key components of PBM, have been the subject of much recent writing on the micropolitics of education and educational research. This writing on dialogue has not, however, been situated within a problem focus, and in the following brief discussion of the work of Gitlin (1990) and Burbules (1991) I want to further explore the significance of such a focus, by considering its implications for their positions. Gitlin (1990) describes educational research as alienating because it "for the most part silences those studied, ignores their personal knowledge, and strengthens the assumption that researchers are *the* producers of knowledge" (p. 444). He builds on this critique to outline an alternative research process, called educative research, which has the development of voice as its central aim. Borrowing from Hirschman's (1970) analysis, Gitlin defines voice as "both an articulation of one's critical opinions and a protest. The protest is not simply a gripe but a challenge to instances of domination and oppression" (p. 459).

While the development of voice is compatible with PBM, it leaves the researcher and practitioner far short of the tools needed to understand and resolve educational problems. Problems of practice may not be detected or correctly understood because voices are suppressed, but problem resolution also requires rational and mutually acceptable ways of silencing voices, so that dissent and plurality, while recognised, does not preclude consensually co-ordinated action. For many educators, appeals for more inclusive dialogue are frightening and only half credible, for they raise the prospect of greater complexity and greater uncertainty, while ignoring the simultaneous need to coalesce voices around one course of action. A problem focus provides a context around which the legitimacy and validity of various voices, once heard, can be consensually agreed. Its emerging constraint structure progressively rules out certain possibilities and rules in others, satisfying the need to both hear the voices of all affected and to reduce their diversity in order to take collaborative action.

Incorporation of a problem focus would also, I believe, advance many of the theoretical debates about the micropolitics of education and educational research that appear irresolvable when conducted in abstract terms. Take, for example, the debate which is lucidly summarised by Burbules (1991), about whether dialogue across difference is possible or even desirable, given that it may involve imposition of a dominant group's values, beliefs and mode of discourse. Burbules explores some of the thorny problems involved in assumptions about intersubjectivity, rationality and consensus on the one hand, and incommensurability, imposition and diversity on the other. He concludes, and I would agree with him, that rather than debate these questions in abstract and extreme form, we reframe them as questions about the conditions under which dialogue across difference is possible, and about the significance and sufficiency of various combinations of understanding and misunderstanding, diversity and unity. Questions about significance and sufficiency, however, cannot be answered in the abstract; they are relative to particular purposes against which the significance and sufficiency of particular levels of unity and diversity, for example, are judged. If such debates were conducted in the context of specific problems, their constraint structure would provide the standards against which these judgements could be made. After pursuing this pragmatic course, we would no doubt discover that the resolution of some problems required tightly co-ordinated action, and therefore high levels of unity and consensus, while the resolution of other problems required less co-ordination, and therefore lower levels of unity and consensus.

Obstacles to Progress with PBM

The claims made in the introductory and concluding sections of research reports suggest that many educational researchers believe their studies are relevant to practice. Lack of desire to make a connection between research and practice is therefore not a major obstacle to the progress of PBM. What is far more likely to be an obstacle is the unwillingness or inability of researchers to make the methodological choices that I have argued are required to achieve this goal. Overcoming this obstacle is important, because until more PBM exemplars become available one cannot judge the effectiveness of the methodology. What would be required, then, for more researchers to adopt this approach to their practice-oriented projects? On the positive side, since many of the constituent features of PBM are drawn from existing research approaches, the resources required do not have to be created afresh. Within the educational and social science research communities are those who have provided superb resources on matters of dialogue and the politics of research, those whose expertise lies in the interpretivist reconstruction of practitioner knowledge, and those whose knowledge of social theory informs their critical empirical work. On the negative side, more researchers are needed who are willing and able to employ the particular combination of resources that I have argued are needed to understand and resolve educational problems. To oversimplify some previous complex arguments, interpretivists and positivistically inclined empiricists lack the normative theorising essential to critique, many empiricists employ social relations of inquiry that are antithetical to mutually educative change, and few of them, in common with critical researchers, embed their inquiry in an inclusive process of problem-solving. If we are to increase the pool of researchers who can use PBM, research training will have to be more responsive to its subject matter, i.e. to the nature of educational problems, and less responsive to the disciplinary and paradigmatic boundaries that currently govern much research training. Normative theorising is essential to researchers who purport to help those whose practice is saturated with normative questions; intervention theory and skills are essential to those who want to translate their emancipatory and educative espousals into educative practice in politically constrained and complex settings. The former is usually taught to those who eschew empirical research training, the latter to those who go down a consulting rather than a research path. Those researchers who want to make a difference cannot afford these distinctions; the problems they want to address require openness to, if not expertise in, all of them.

Since PBM is built around a process of critical dialogue with all those involved in a focus problem, it suggests that research-driven educational improvement requires intensive face-to-face contact between researchers and practitioners. Such an expensive and time-consuming process could be a severe constraint on the influence of PBM. This problem could be overcome, in some cases, if research reports incorporated some of the key features of critical dialogue. By this I mean that they would describe generalisable features of practitioner theorising about particular types of problem, in ways that had face validity for readers, and would critique such theorising in a manner that was intellectually compelling. Neither will be achieved if researchers continue to investigate educational problems in ways that bypass practitioner theories, or if critique is couched in a language that is inaccessible. Assuming that practitioners can locate their own situation in the report, and that they find its critique intellectually compelling, the problem of translating intellectual conviction into changed practice remains. Practitioners could change themselves, either individually or collectively, if the alternative practice suggested by the report could be adopted through a process of adjustment or single-loop learning. If more fundamental double-loop change was required, this would be far more likely to require an intensive face-to-face educative process, because double-loop changes require us to recognise and publicly re-examine our taken-for-granted assumptions about the world. This type of learning, given the strength of the emotional, cognitive and institutional barriers to inquiry into many educational problems, is highly unlikely to be accomplished through reading and reflecting on a research report or participating in a day's in-service education. Solving these problems requires that practitioners unlearn the processes that have so far prevented them from doing so, and learn how to practice the skills of critical dialogue around issues that were previously bracketed out of such inquiry. This is the type of learning and the type of problem-solving that requires an intensive face-to-face dialogue with a researcher or interventionist who is skilled in these processes. The future contribution of educational research to the resolution of this type of problem is critically dependent on the dialogical competence of researchers themselves and of those who teach them.

References

Burbules, N. C. (1991). Dialogue across differences: Continuing the conversation. *Harvard Educational Review*, **61** (4), 393–416.

Gitlin, A. D. (1990). Educative research, voice and school change. *Harvard Educational Review*, **60** (4), 443–466.

Hirschman, A. (1970). *Exit, voice and loyalty*. Cambridge: Harvard University Press.

Strike, K. (1989). *Liberal justice and the Marxist critique of education: A study of conflicting research programs*. New York: Routledge.

Young, R. (1989). *A critical theory of education: Habermas and our children's future*. New York: Harvester Wheatsheaf.

APPENDIX A
Western College Staff Appraisal Policy Draft

Rationale

Staff appraisal is an integral aspect of professional development linked as it is to the continuous process of improving and extending professional skills.

Every professional organisation must engage in professional development in one form or another in order to credibly survive. If we accept that the principal purpose of appraisal is to improve the quality of student learning, then the student interests are seen to coincide closely with the professional development needs of teachers and administrators.

Appraisal is thus viewed as a professional development interim and interventional strategy—both retrospective and prospective.

It involves:

looking back on what has been achieved,
looking back on what has not been achieved,
taking stock of the present,
planning strategies which will assist development in the future.

Purpose

Professional development. To ensure that all staff are given opportunities to improve and enhance their professional skills. In order to achieve this purpose staff responsible for others must be knowledgeable about their practice, professional development needs, professional achievements, ability and potential. Opportunity for reciprocal appraisal makes it possible for all staff to obtain critical feedback from colleagues.

Improving classroom skills. To improve the quality of student learning through classroom observation and analysis of classroom processes. Identifying strengths and weaknesses and planning intervention strategies which will help the teacher develop further in the future. The HOD as instructional leader is able to contribute to professional development initiatives and negotiate staff and curriculum development and resource allocation for meeting these needs.

Improving management skills. Clarification and negotiation of roles and responsibilities underpinned by acceptance of accountability for the professional tasks undertaken in a position or area of responsibility, e.g. instructional leadership; managing staffing, curriculum, resources; implementation of school policies and programmes. Executive staff as senior administrators contribute to the co-ordination of professional development need identification and resourcing.

SCHOOL DEVELOPMENT

Staffing needs. Deployment, classification and promotion.

GUIDELINES

1. The attached MODEL FOR STAFF AND SCHOOL APPRAISAL outlines responsibilities for and the reciprocal nature of the appraisal model suggested for the school.
2. Structures which provide for the formal and regular appraisal of staff should be applied universally using a similar process.
3. Appraisal should be OPEN, RECIPROCAL and POSITIVE.

   Open: Saying all the things that are important to us, relevant to our agreed agenda in a way that allows others to express different views.

   Reciprocal: Both the appraiser and the appraisee contribute to the judgements that are made, and have the opportunity to influence one another.

   Positive: The appraisal process is directed towards helping individuals improve their performance regardless of its current level in ways that are mutually agreed.

(**General Comments on Skill:** The ability to apply these principles, especially in the context of giving and receiving difficult feedback, will require all of us to learn and develop special communication skills.)

4. Issues of confidentiality, trust, values and attitudes should be discussed in a climate of professional and collegial honesty at all levels of the school.
5. Self-appraisal questionnaires should be designed by HODs and executive staff.

Conclusion

There is no doubt that critical review and judgement must be employed in the process of appraisal, nor can those responsible for the performance of others avoid the professional responsibility for appraising. If the intention and the process are open, honest and professional then the staff involved at all levels of the institution can be prepared to expect and accept critical appraisal as a two-way reciprocal process. It is a process that should be approached positively in an effort to give and take supervision, support, observation, analysis, feedback, guidance and judgement which will impact effectively on the learning–teaching activities in the school.

Without staff appraisal professional development does not effectively target individual and organisational needs.

A MODEL FOR STAFF AND SCHOOL APPRAISAL

Ministry of Education
Review & Audit Agency
(Appraise the school in relation to the objectives stated in the CHARTER)

↑ ↓

Board of Trustees
(Appraise the principal in terms of responsibility for implementing the Charter)

↑ ↓

Principal
(Appraises staff and systems as educational and organisational leader of the institution)

↑ ↓

Executive Staff
(Appraise the senior and middle management group in terms of their management responsibilities. They also appraise the principal and all staff generally)

↑ ↓

Middle & Senior Management Staff
(Appraise staff for whom they are responsible and also appraise executive staff)

↑ ↓

Staff
(Appraise the efforts of those responsible for their appraisal and have access to HOD, collegial and self-appraisal as on-going professional development strategies)

The School Development Process which incorporates collaborative decision-making and problem-solving allows opportunities for on-going school appraisal as an element of staff professional development.

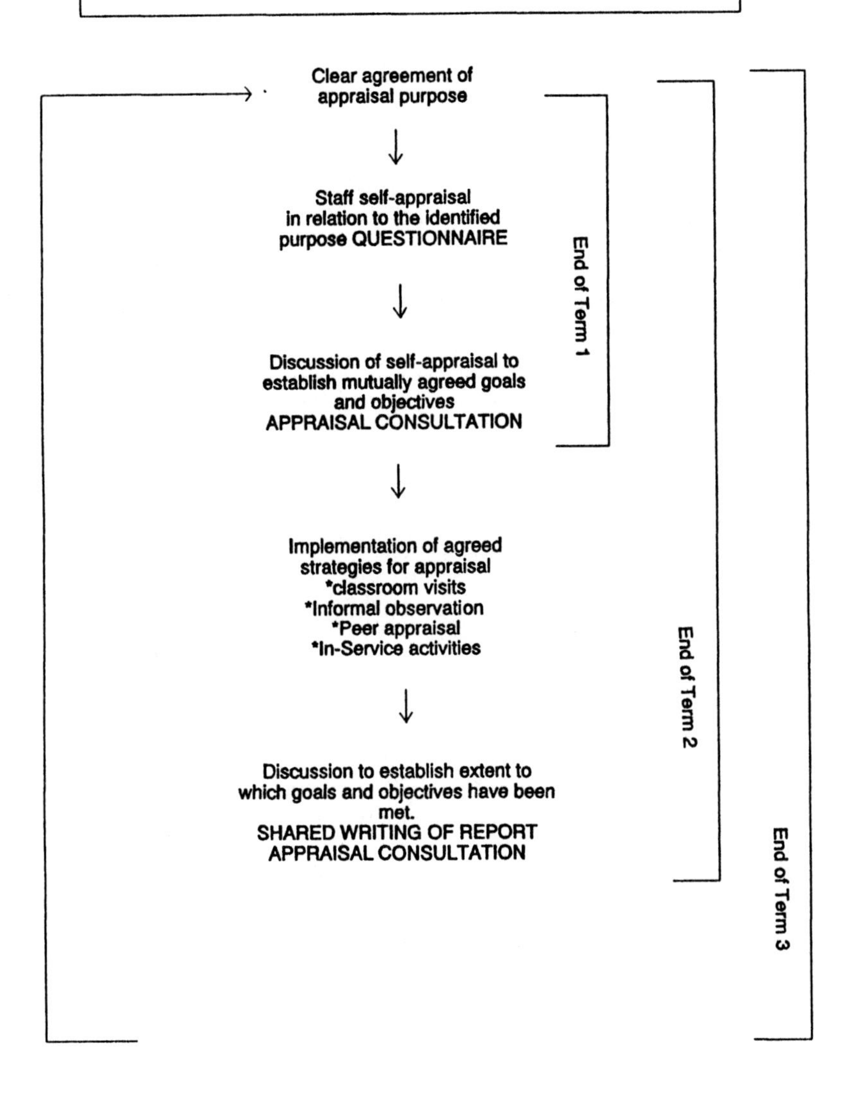

Executive Staff → Single Staff Departments
Head Of Departments → Staff in Department
Clear agreement of appraisal purpose
Staff self-appraisal in relation to the identified purpose QUESTIONNAIRE
Discussion of self-appraisal to establish mutually agreed goals and objectives
APPRAISAL CONSULTATION
Implementation of agreed strategies for appraisal
*classroom visits
*Informal observation
*Peer appraisal
*In-Service activities
Discussion to establish extent to which goals and objectives have been met.
SHARED WRITING OF REPORT
APPRAISAL CONSULTATION
End of Term 1
End of Term 2
End of Term 3

APPENDIX B
Western College Staff Appraisal Policy

Writers: JH/EL/ML
Date: September 1989

Rationale

Staff appraisal is an integral part of professional development. It is linked to the process of improving and extending professional skills and is a strategy to facilitate this. The process of appraisal looks back on what has and has not been achieved, takes stock of the present and plans for future development.

Purposes

1. To facilitate better student learning by improving the quality of classroom practice.
2. To support staff in developing their professional skills.
3. To ensure that all staff are given opportunities to enhance their professional skills.
4. To identify their strengths, needs and potential.
5. To improve management skills.
6. To clarify the roles and responsibilities of teaching staff.
7. To identify targets for school development.
8. To encourage team building.

Guidelines

1. Appraisal should be open, reciprocal and positive.
2. Structures providing for the formal and regular appraisal of staff should be applied universally using a similar process.

3. Each teacher should have at least one appraisal consultation each term.
4. The first interview of the appraisal cycle should include a review of the job description and the setting of goals.
5. Goals should be positive and achievable.
6. Once goals are set, forms of support should be identified.
7. Consultations should focus on classroom/management practice, based on an agreed information gathering programme which includes visits to classrooms and departmental meetings.
8. During the cycle, a self-appraisal questionnaire should be completed.
9. The final consultation of the appraisal cycle should result in a mutually agreed report, derived from the self-appraisal questionnaire, an evaluation of progress towards meeting the year's goals and of management support.
10. Assistant teachers consultations are with the HOD with whom they have a functional relationship.
11. HODs' consultations are with the Executive staff member with whom they have a functional relationship.
12. Senior staff meet with the Principal.
13. Where a staff member feels concern with the allocation they should feel free to raise the matter.
14. The provision of training in the skills of appraisal is a vital, ongoing part of professional development.
15. The formal appraisal consultation cycle does not preclude recognition of the importance of informal contact/consultation.
16. All staff must recognise the importance of confidentiality and trust.
17. Consultation should take place in a climate of professional and collegial honesty.

Conclusion

Without staff appraisal, professional development does not effectively target individual and organisational needs.

There is no doubt that critical review and judgement must be employed in the process of appraisal. If the intention and the process are open, honest and professional then all staff can be prepared to expect and accept appraisal as a two-way reciprocal process.

It is a process that should be approached positively and it should then create a climate in which sympathetic collegial support will assist individuals to achieve constructive personal development.

Staff concerns about the nature and purpose of appraisal should be addressed by ensuring that the process is open and honest, that it is based on agreed criteria and that every effort is made to meet staff professional needs.

Recommendations

1. The role of the HOD should be clarified and specified.
2. Priority should be given to the provision of a suitable office for consultations.

Author Index

Subject Index

Printed in the United Kingdom
by Lightning Source UK Ltd.
120317UK00001B/42